Advances in Electronic and Electrochemical Ceramics

Advances in Electronic and Electrochemical Ceramics

Ceramic Transactions Volume 179

Proceedings of the 107th Annual Meeting of The American Ceramic Society, Baltimore, Maryland, USA (2005)

Editors
Fatih Dogan
Prashant Kumta

Published by
The American Ceramic Society
735 Ceramic Place, Suite 100
Westerville, Ohio 43081
www.ceramics.org

Advances in Electronic and Electrochemical Ceramics

For information on ordering titles published by The American Ceramic Society, or to request a publications catalog, please call 614-794-5890, or visit www.ceramics.org

ISBN: 1-57498-262-1
ISBN 13: 978-1-57498-262-6

10 09 08 07 06 5 4 3 2 1

Contents

Preface

Electronic ceramics in extreme environments, e.g., high temperatures and high electrical fields, must operate reliably and efficiently. Such electronic components will not only provide tolerance to hostile environments, but also will reduce system size and weight by eliminating radiators, improving reliability and lifetime, and increasing energy densities.

There is an increasing demand for new advanced functional electronic ceramics in the fuel cell area as well. Fuel cells are gaining considerable momentum in recent years. The growing environmental concerns and the exponential growth in portable electronic devices has created intense worldwide activity on fuel cells. Fuel cells offer clean energy and are attractive for a variety of power needs ranging from electric vehicles to stationary power and even handheld devices such as cellular and mobile phones.

The 107th Annual Meeting & Exposition of The American Ceramic Society took place in Baltimore, MD, April 10-13, 2005. This volume documents a special collection of articles from a select group of prominent scientists from academia and industry who presented their work at this meeting and primarily in the Electronic Ceramics in Extreme Environments and Fuel Cells and Related Systems symposia. These articles represent a summary of the oral presentations focusing on both the scientific and technological aspects of electronic ceramics in extreme environments and fuel cells. This select collection of manuscripts thus provides pertinent and state-of-the-art information from prominent scientists and engineers from academia, national laboratories, and industry on the latest developments in the above areas.

The success of the symposia and the issuance of the proceedings could not have been possible without the support of staff at The American Ceramic Society Headquarters and the other symposia co-organizers Drs. Ming Jen Pan, Ali Sayir, Bruce Tuttle, Arumugam (Ram) Manthiram, JinYong Kim, Vincent L. Sprenkle and Ki Hyun Yoon. Financial support from The American Ceramic Society and Sandia National Laboratories is gratefully acknowledged. The organizers are grateful to all participants and session chairs for their time and effort, to authors for their timely submissions and revisions of the manuscripts, and to reviewers for their valuable comments and suggestions.

Fatih Dogan, University of Missouri-Rolla
Prashant Kumta, Carnegie Mellon University

Electronic Ceramics for Extreme Environments

EXTREME ENVIRONMENT POTENTIAL OF DIAMOND DERIVED DEVICES

R. S. Takalkar, P. Hamari , J. L. Davidson, W. P. Kang, , P. Taylor, Y. M. Wong
Department of Electrical Engineering and Computer Science
Vanderbilt University, Nashville, TN 37235, USA
jld@vuse.vanderbilt.edu

ABSTRACT

Diamond is an excellent material from which to fabricate field emission devices due to its low electron affinity, high mechanical hardness, chemical inertness and highest thermal conductivity. Diamond D-VFETs (diamond vacuum field effect transistors) offer an interesting alternative to "beyond silicon" solid state semiconductor devices. They can operate unchanged in a variety of harsh environments including radiation and extreme temperatures. We have developed micro-patterned diamond pyramidal tips and edge emitters. In this paper, we report the high temperature behavior of these diamond emitters. The polycrystalline diamond films were grown by plasma-enhanced chemical vapor deposition. The effect of high temperature was investigated by measuring current-voltage from test devices at various temperatures. The turn-on voltage of diode type configurations for both types of emitter geometries (tips and edges) was not affected by increasing temperature. The emission current and forward and reverse characteristics were unaffected by temperature to > 400 °C. Fowler-Nordheim emission was confirmed for all temperatures.

Diamond power resistors have the advantage of rapid dissipation of Joule heating, assured by the high thermal conductivity of the material and accommodating temperature cycling (extremely small TCE). Thin film diamond resistors of various sizes were fabricated on insulative AlN substrates, boron doped with trimethylboron (TMB) gas and patterned by direct reactive ion etching. Resistors with different device geometry and doping concentration were examined for power density response. Arrays of microstructure resistors were exposed to pulsed voltage to examine their thermal response.

INTRODUCTION

The unique properties of diamond such as low or negative electron affinity [1], excellent mechanical properties such as high hardness and toughness and highest thermal conductivity and wide band-gap make it an excellent material for applications with extreme ambient conditions. In this article we present application of diamond to fabricate vacuum field emission devices and power resistors capable of operating without performance degradation beyond current Si based soilid state devices.

Electron field emission from diamond has been observed experimentally to yield high emission current at low electric fields [2-5]. With no solid state semiconductor properties, electron field emission from diamond is expected to be temperature insensitive over a much wider temperature range than conventional silicon devices, for operation in harsh environments. Diamond resistors have ohmic behavior at low to medium current levels and then begin to experience thermal excitation (joule heating), Carrier density enhancement and hence conductivity increases at high current levels. Diamond's tolerance for high current density allows the resistors to continue to operate at high power and temperature conditions after entering a thermal 'runaway' situation.

FABRICATION OF DIAMOND DEVICES

Diamond Vacuum Field Effect Transistors

The fabrication steps for diamond pyramidal tips and edge emitters are shown in Fig. 1. The Si wafer was 500μm thick. A 0.2μm thick oxide was then grown on the wafer surfaces. Inverted pyramidal cavities were then etched into the Si substrate using photolithographic patterning and anisotropic etching of silicon with KOH solution. Next a SiO_2 layer was grown into the mold to produce a well sharpened apex in the inverted pyramidal cavity. Diamond was then grown into the mold using plasma enhanced chemical vapor deposition (PECVD). The PECVD parameters were controlled so as to achieve small but deliberate sp^2 content in the diamond film. Next the back-side silicon and SiO_2 was etched away and sharpened diamond pyramidal geometries exposed. The diamond film was characterized using scanning electron microscopy (SEM) and Raman spectroscopy. The SEM image of an array of diamond pyramidal tips and edge emitters is shown in Fig. 2 and 3 respectively. The tips had a base dimension of 12μmX12μm, while the edges had a width of 2μm and length of 125μm.

Emission testing was carried out under vacuum at 10^{-6} Torr. The sample, in a diode configuration, was placed on top of a button heater to perform emission measurements at elevated temperatures. A mica spacer 120μm thick provided the cathode-anode gap. Emission measurements at various temperatures were performed after the vacuum stabilized at 10^{-6} Torr. The current-voltage behavior was recorded to a computer interfaced with the test chamber.

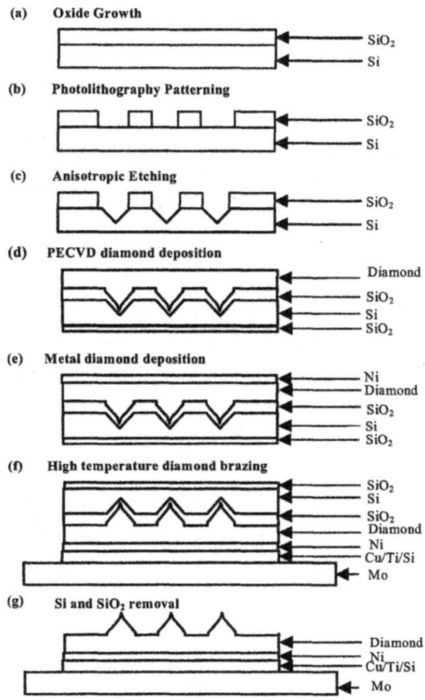

Fig. 1 Fabrication sequence of diamond field emitter devices.

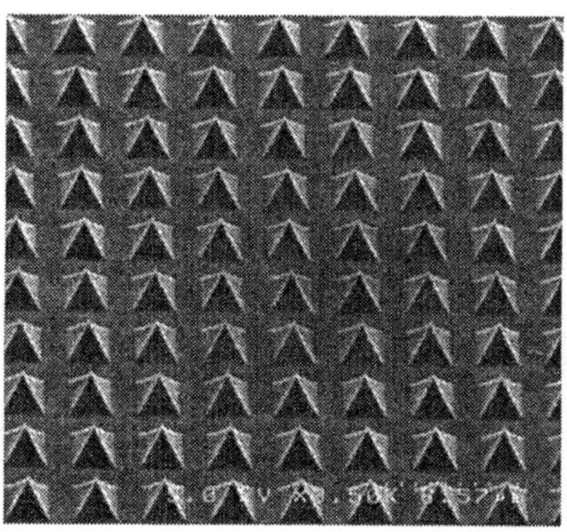

Fig. 2 SEM of micropatterned diamond pyramidal tips.

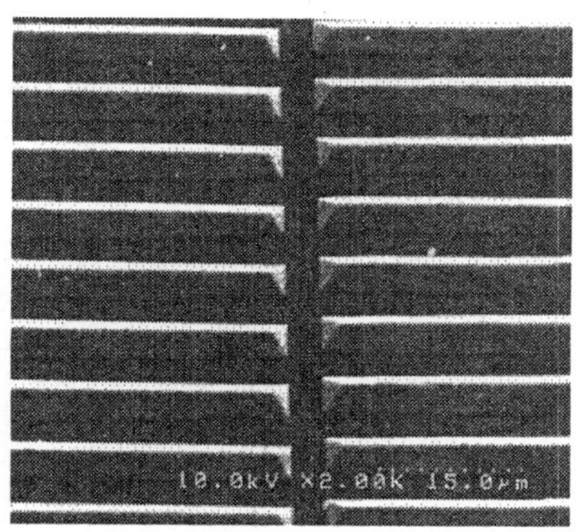

Fig. 3 SEM of micropatterned diamond Edge emitters

Diamond Power Resistors

Polycrystalline diamond films of approximately 2 microns in thickness were deposited on an aluminum nitride substrate by a PECVD process. Boron doping was from a trimethylboron (TMB) gas source. The PECVD diamond film was deposited on the surface substrate in a chamber with an atmosphere that was a mixture of hydrogen (H_2), methane (CH_4), and TMB for 2.5 hours. The gases were introduced at flow rates of 478 sccm, 18 sccm, and 10 sccm respectively. The substrate temperature was 800°C at a dynamically pumped pressure of 120 Torr. The plasma was sustained by a microwave source operating at 5 kW and 2.45 GHz.

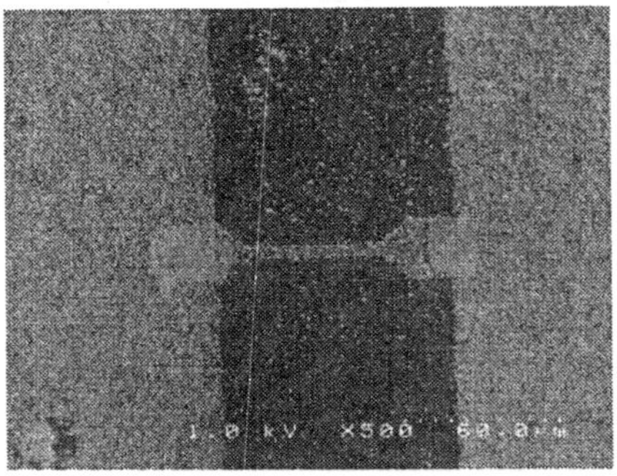

Fig. 3 Low magnification SEM of diamond resistor on AlN substrate (500 x).

Aluminum (Al) metal contacts, 1μm thick, were deposited on the resistors. The resistor geometries were formed by direct etching in oxygen plasma from the CVD diamond layer using a patterned metal mask. Titanium was chosen as the masking metal because of the good masking properties that have been exhibited with titanium in previous experiments [6].

Four different resistor structures were fabricated for this study. The resistors were "necked" in shape, as can be seen in Fig. 3, and ranged in length from 50 μm to 11 mm. The necking pattern is designed to enhance and concentrate current flow. The structures were 10μm to 90μm in width. A closer look at the topology is illustrated in Fig. 4. Overall, this particular family of resistors physically the smallest set to be experimentally examined.

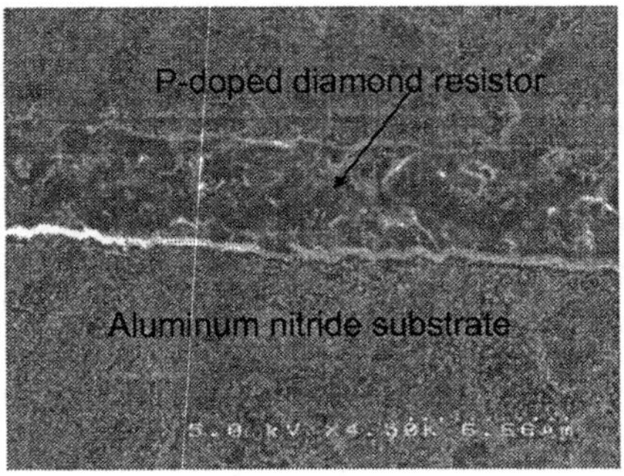

Fig. 4 High magnification SEM of resistor neck on AlN substrate (4500 x).

The Fowler-Nordheim (F-N) relation was used to describe the field emission data, via;

$$\ln(I/E^2) = \ln(A{*}K1{*}\beta^2/\Phi) - (K2{*}\Phi^{1.5}/\beta)(1/E) \quad (1)$$

where K1 and K2 are constants : $K1 = 1.54 \times 10^{-6}$ (AeV/V^2), $K2 = 6.83 \times 10^7$ V/(cm $eV^{3/2}$), I is the emission current, Φ is the work function of the emitting surface in eV, β is the geometrical field enhancement factor, A is the emitting area, and E is the applied electric field.

Fig. 6 and 7 show the field emission behavior of diamond pyramidal tip emitter at various temperatures and its corresponding F-N plot. The emission current was observed to be insensitive to change in temperature up to 400°C. A turn-on field of 6V/μm was observed and it did not change with temperature. The straight line F-N plot in Fig. 7 demonstrates that the emission current of the diamond diode conforms to F-N behavior. Fig. 8 shows the emission current vs. temperature at various applied electric field and demonstrates the temperature independence of the field emission current. The diode also demonstrates excellent reverse characteristics with immeasurable leakage current (<1pA) and a high break-down voltage, which is limited by mica breakdown. The diode reverse characteristics was unaffected by temperature changes.

The field emission behavior of diamond edge emitter at different temperatures is shown in Fig. 9 and the corresponding F-N plot in Fig. 10. The emission current again shows insensitivity to temperature changes up to 300°C. The edge emitters show a turn-on voltage of 7V/μm. The straight line F-N plot in fig. 10 demonstrates that the emission current of this diamond diode conforms to F-N behavior. Fig. 11 shows the emission current vs. temperature characteristic at various applied electric fields. The edge emitter diode reverse characteristics are similar to that observed in tip emitters.

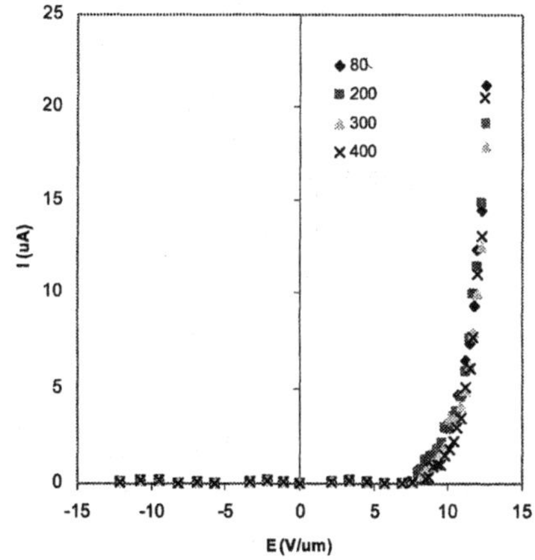

Fig. 6 I-E plot of diamond tips for various temperatures.

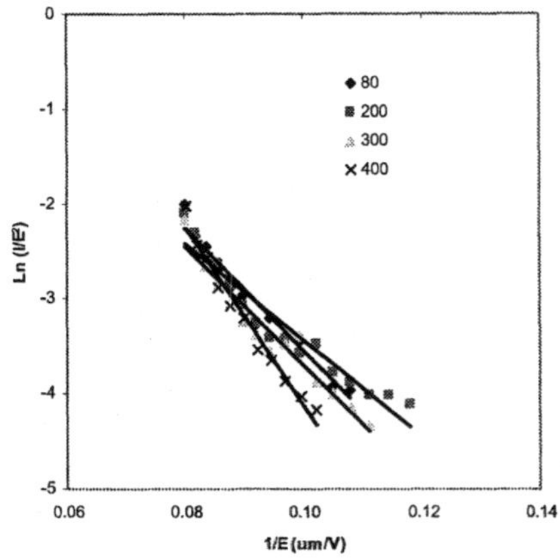

Fig. 7 F-N plot of field emission from diamond tips for various temperatures.

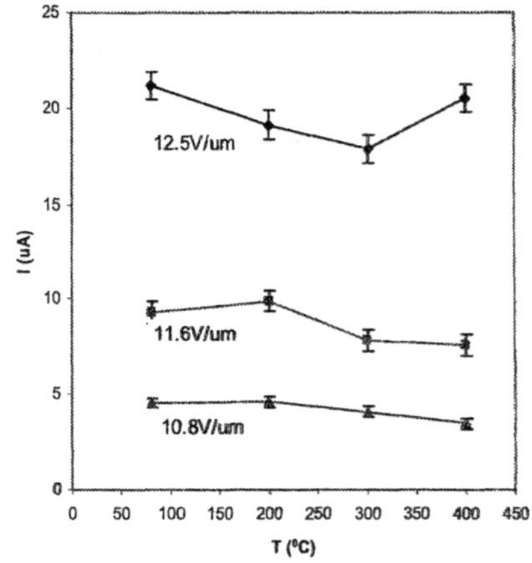

Fig. 8 Current-temperature plot of field emission from diamond tips.

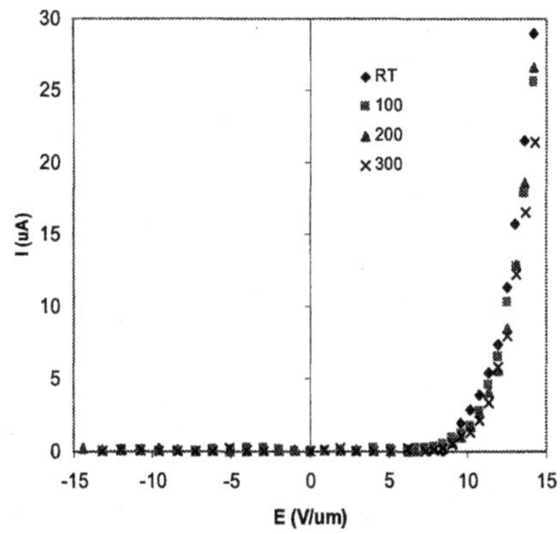

Fig. 9 I-E plot of diamond edge emitter for various temperatures.

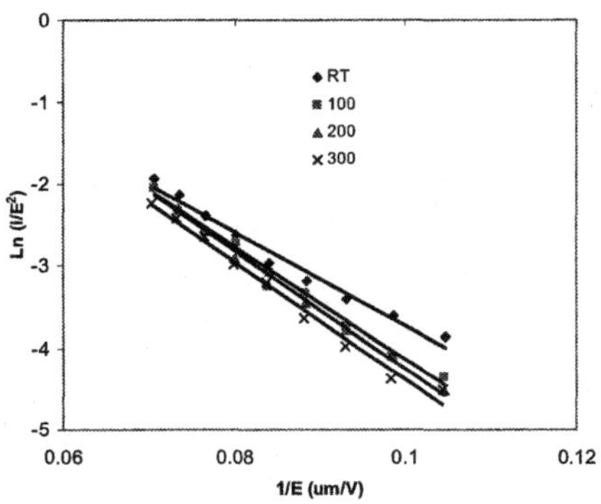

Fig. 10 F-N plot of field emission from diamond edge emitters for various temperatures.

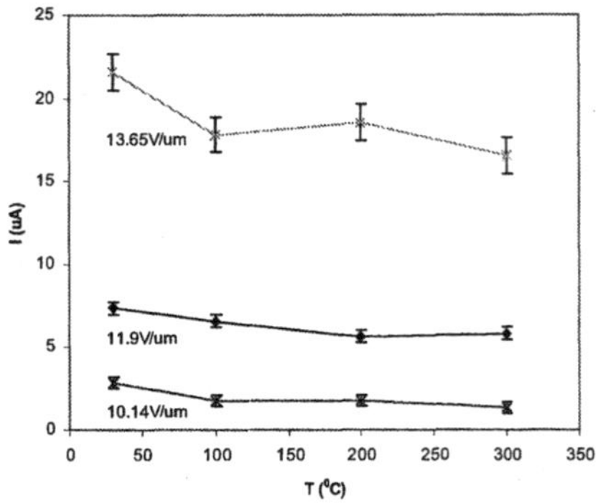

Fig. 11 Current-temperature plot of field emission from diamond edges.

The temperature insensitivity of diamond electron field emission devices is unlike that of solid state devices, which are temperature sensitive. The observed temperature independence of diamond field emission devices can be explained as follows. In the Fowler-Nordheim equation used to analyze field emission, the only variable that is temperature dependent is the work function (Φ). Diamond is a wide band-gap material with a work function of 5.45eV. Considerable thermal energy is needed to produce an electron-hole pair across such a big band-gap. Hence, the intrinsic carrier concentration (n_i) in diamond remains unaffected, unless it is heated to a high temperature. The maximum temperature insensitivity of diamond can be determined by examining it for thermionic emission, a subject of another study. Silicon junction device, as well known, will become "soft" when at 100°C. The limit for high temperature of diamond vacuum devices is undetermined. The high temperature limit of 400°C during this work was determined by the mica spacer leaking electrically above this temperature.

Diamond Power Resistors

Diamond resistors were fabricated on a ceramic substrate of aluminum nitride to investigate the power density characteristics of p-doped (boron) diamond.

This study was intended to examine the high power density characteristics of p-doped (boron) diamond. The resistors had ohmic behavior at low-to-medium current levels and then began to experience thermal excitation (joule heating). Carrier density enhancement and conductivity increased at high current levels. Diamond's tolerance for high power allowed the resistors to continue to operate at high power and temperature conditions after entering a thermal 'runaway' situation. Operating the resistors under load at high voltage levels is very similar to operating devices in the 'reverse breakdown' region except with much higher power densities.

Resistors were subjected to electrical and thermal stress for evaluation and characterization. Fig. 12 shows the current-voltage behavior of a typical sample. As expected and was observed in previous tests, the resistor behaves in an ohmic fashion at low to medium current levels. As the load increases, the effects of joule heating, carrier density enhancement, and conductivity also increase. This leads to a thermal 'runaway' scenario at high current levels. Diamond resistors continue to under these high stress conditions due to the robust nature of the material itself.

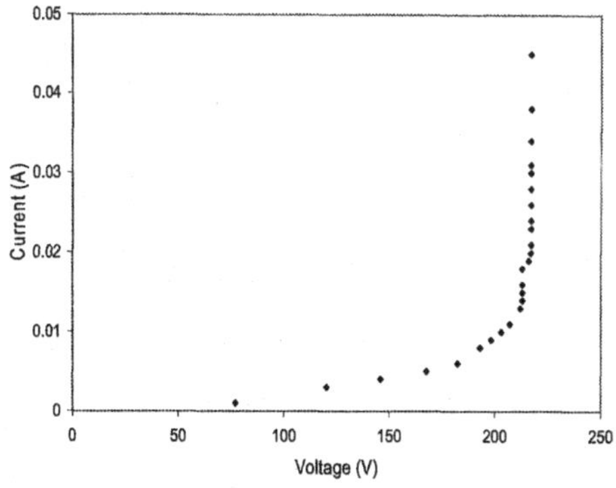

Fig. 12 I-V behavior of sample resistor

The strength of diamond coupled with the small size of these resistors lead to impressive power handling capabilities. The size also leaves these specimens vulnerable to the effects of localized electrical arcing at increased voltages. Since the electrical contacts are only separated by a number of microns, the probability is much greater for air ionization at medium to high voltages. Fig. 13 shows a sample after current loading when a high power electrical arc caused catastrophic damage. To lessen the effects of the arcing phenomena, samples were additionally tested under vacuum conditions.

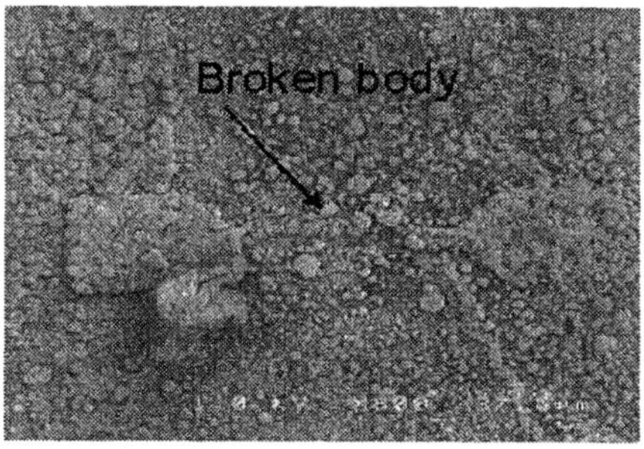

Fig. 13 SEM of damage due to localized electrical arcing.

The current and power densities of this family of power resistors were much higher than those from earlier work. For example, a sample sustained 24.6 W. The cross sectional area of the sample was 1.98 x 10^{-6} cm^2. This yields a power density of 12.4 MW/cm^2. This is superior to the power density of 1.7 MW/cm^2 previously observed. In rough comparison [6], a light bulb filament operates at ~50 kW /cm^2. The average power sustained by a typical resistor was approximately 13 W. This is more than double the 5 W performance seen in previous experiments [7].

Current densities were observed as high as ~9 x 10^4 A/cm^2. The thermal analysis of the resistors also yielded interesting results. Fig. 14 shows the resistance-current relationship of a typical sample. As expected for diamond, there is a sharp decline in the resistance as the current increases. The resistance declines in conjunction with the associated increase in temperature.

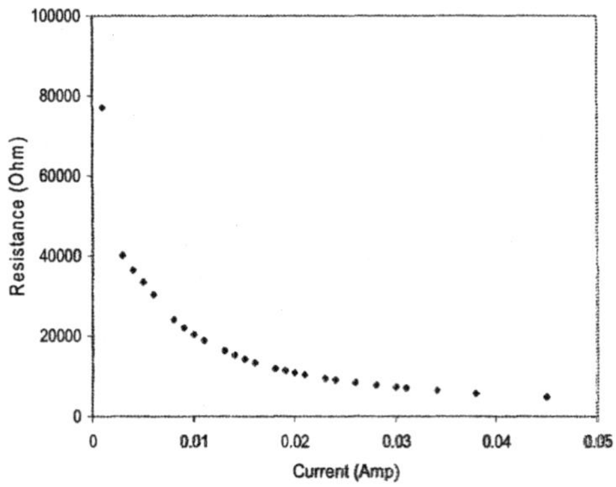

Fig. 14 Resistance vs. current plot for example resistor.

Given that the resistance behavior of boron doped diamond has been observed by many to be Arrhenius in form and that representative samples of these resistors were similarly characterized, the temperature profile for the resistors can be estimated from the power tests [7]. The dopant activation energy for these devices was determined from resistance (R) vs. temperature (T) tests to be 0.095 eV. Fig. 15 illustrates the activation energy determination.

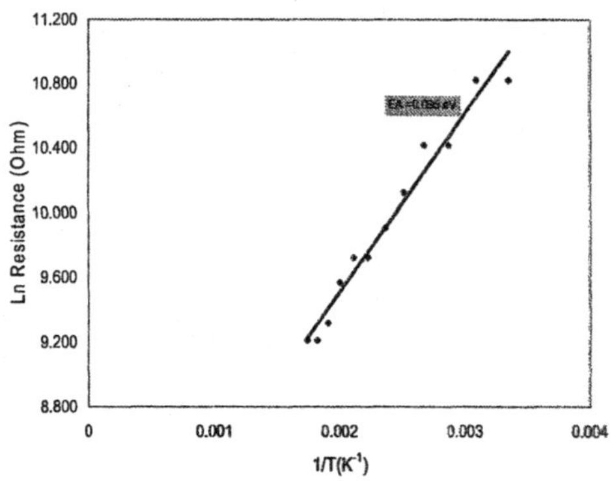

Fig. 15 Determination of activation energy

Fig. 16 SEM of resistor after thermal stress

The use of vacuum testing for this round of samples led to less instances of sample destruction than in previous trials. At high current the resistors experienced enough resistive heating to cause visual illumination in the form of a "red hot" glow. Radiant emission was achieved at temperatures of ~ 400°C. This threshold temperature corresponded to applied voltages of ~200 to ~500 V. Fig. 16 and 17 show the condition of a typical resistor after it was loaded to the point of radiant emission.

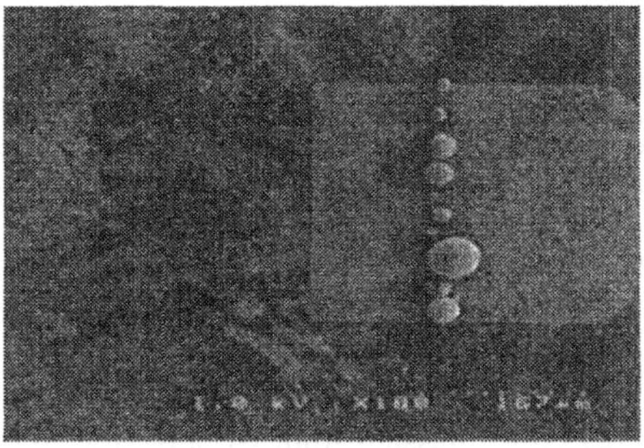

Fig. 17 SEM close-up of contact electrode.

The diamond layer of the resistor was left undamaged. The bubbles on the contacts illustrate the fact that the aluminum contacts had begun to melt under the intense heat of the resistive heating. An Energy Dispersive Spectrometry (EDS) analysis was performed to ascertain the amount of aluminum that had migrated into the resistor body. Conductive particles in the resistor body itself is, another mechanism by which resistors of this nature can be inhibited or brought to failure. As shown in fig. 18, aluminum migration can indeed be found in the sample.

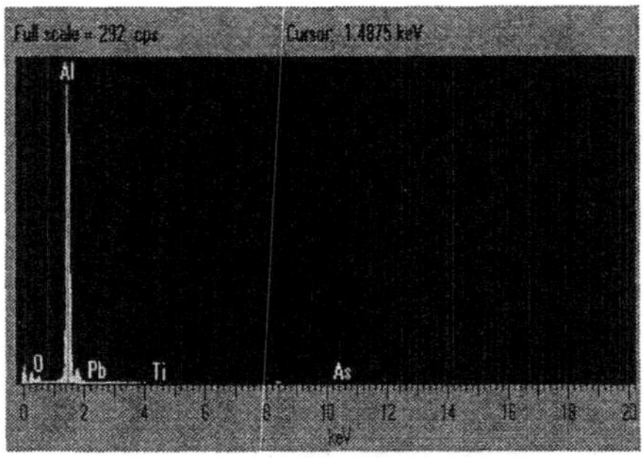

Fig. 18 EDS analysis of resistor after heating

CONCLUSIONS

Diamond Vacuum Field Effect Transistors

In conclusion, we have investigated the effect of temperature on diamond field emission behavior. It was found that, field emission current from diamond is unaffected by temperature up to 400°C. The field emission turn-on voltage is also not affected by changes in temperature. Therefore, diamond field emitters are very well suited for high temperature applications.

Diamond Power Resistors

This new family of p-doped diamond resistors shows excellent promise as a high stress resistive material. The high power densities and extreme temperature handling capabilities are superior to those observed in earlier resistor development. With additional evaluation of temperature parameters through pulse heating trials and further development of contact patterning and development, more robust and reliable structures will be developed and characterized in future experiments.

REFERENCES

[1]C. Bandis, B. Pate, "Photoelectric Emission from Negative-Electron-Affinity Diamond (111) Surfaces: Exciton Breakup versus Conduction-and Emission", Physical Review B, **Vol. 52 No. 16**, 12056-71, (1995).
[2]V. Zhirnov, E. Givargizov and P. Plekhanov, "Field Emission from Silicon Spikes with Diamond Coatings", Journal of Vacuum Science and Technology B, **Vol. 13, No.2**, 418-21, (1995).
[3]E. Givargizov, "Silicon Tips with Diamond Particles on them: New Field Emitters?", Journal of Vacuum Science & Technology B, **Vol. 13, No. 2**, 414-17, (1995).
[4]J. Liu, V. Zhirnov, G. Wojak, A. Myers, W. Choi, J. Hren, S. Wolter, M. McClure, B. Stoner, J. Glass, "Electron Emission from Diamond Coated Silicon Field Emitters", Applied Physics Letters, **Vol. 65, No. 22**, 2842-44, (1994).
[5]T. Asano, Y. Obuchi, S. Katsumat, "Field Emission from Ion-milled Diamond Films on Si", Journal of Vacuum Science and Technology B, **Vol. 13, No. 2**, 431-34, (1995).

[6]J. Davidson, W. Kang, K. Holmes, A. Wisitsora-At, P. Taylor, V. Pulugurta, R. Venkatasubramanian and F. Wells. "CVD Diamond for Components and Emitters", Dia. And Relat. Mater. **10**, 1736-42, (2001).
[7]K. Holmes, "Piezoresistive Micro-ElectroMechanical Applications in Diamond Films", Doctoral Thesis Dissertation, Vanderbilt University, December 2002.

DIELECTRIC POWDER/POLYMER COMPOSITES FOR HIGH ENERGY DENSITY CAPACITORS

Lynell J. Gilbert, Thomas P. Schuman
University of MO-Rolla - Chemistry
142 Schrenk Hall
Rolla, Mo, 65409

Fatih Dogan
University of MO-Rolla –Materials Science and Engineering
222 McNutt Hall
Rolla, Mo, 65409

ABSTRACT

In this work, we present the synthesis, characterization, and electrical performance of an epoxy/ barium titanate composite. First, the $BaTiO_3$ ceramic powder was examined by X-ray diffraction and scanning electron microscopy and was found to be tetragonal in its crystal structure with a particle size of approximately 1µm and crystals of 57.7nm. It was then established the particles of $BaTiO_3$ could be well dispersed into an epoxy resin matrix using the common ceramic dispersant Menhaden fish oil, in 80/20 ratio by weight xylene to ethanol solvents, via ball milling. Films were synthesized in the geometry specified by ASTM D2149-97, and a dielectric constant of 13 and loss of 0.018 were achieved for composites with a volume filler concentration of 25%. The breakdown strength was 1243 kV/cm at a volume filler concentration of 7 % and decreased as the concentration of $BaTiO_3$ was increased. Energy densities were calculated assuming a linearly behaved composite and were found to be 0.26-0.36 J/cm^3 for specimens filled with 7 to 25 % barium titanate.

INTRODUCTION

As the consumer desire for faster, more powerful, and smaller electronic devices increases, the design of more effective passive components, such as capacitors, is required because of their current occupation of a large substrate surface area. A transition to integral capacitors, namely ceramic-polymer composites, has been advanced through the examination of composites utilizing ceramic materials in a range of particle filler concentrations and sizes from micro to nanometer.[1-7] Several aspects of capacitor design remain to be considered, however, for a composite to fulfill the role of a high energy storage device: dielectric constant, loss, breakdown strength, flexibility, and ease of processing.

Traditionally, sintered ceramics have been used as surface mounted capacitors because of their high dielectric constants and low loss. Difficulties arise with the exclusive utilization of a ceramic as an imbedded capacitor for printed wire boards because of the difficulty of manufacture and the brittle nature of the fired material. The incorporation of polymers has potential in aiding to overcome these problems, but due to their low dielectric constant must be used in conjunction with a higher κ substance. The electrical properties of a hybrid material will be dictated by the components' intrinsic behaviors, and their interactions with one another.

Because of the ease of processibility, including room temperature cure and achievable flexibility, epoxy has often been utilized as a host matrix.[1-7] In an effort to achieve high dielectric constant composites, several studies have been performed with the incorporation of both lead magnesium niobate- lead titanate (PMN-PT) and barium titanate ($BaTiO_3$) in a range of filler volumes because of their high permitivities, 19,316 and 3,000, respectively.[3-6] A κ of 150 at 85 % volume loading has been achieved by Rao, Yue, and Wong using the two fillers; however, the substantial loading produced problems in dispersion of the ceramic fillers within the matrix. Additionally, there were reported a lack of adhesion and breakdown voltages of the composite were measured at 170 kV/cm.[6]

During the pursuit of a high dielectric composite, little information has been published with regards to a systematic study of all the aspects vital to the creation of a high-energy storage device. It has been established the dielectric constant of $BaTiO_3$ strongly depends on the particle size, grain size, and crystal structure, whether tetragonal or cubic.[8,9] When grain size is approximately 1 μm, the κ of $BaTiO_3$ is at a maximum and decreases as size decreases.[9] The objective of this paper was to study the low volume concentration ($0<\varphi<25$) incorporation of barium titanate particles into an commercial epoxy- polyamide matrix via dispersion with Menhaden fish oil. The effect of particle size, crystal structure, and dispersion will be measured by X-ray diffraction, scanning electron microscopy, and electrical measurements, such as capacitance and loss at 1kHz to 10MHz, will be performed. The relationship between capacitance C and dielectric constant of the composite ε_r is given by the following equation where A is the area, t is thickness, and ε_o is the permittivity of free space.

$$C = \frac{\varepsilon_o \varepsilon_r A}{t} \quad (1)$$

Finally, the breakdown strength will be determined and energy density calculated using the relationship for a linear dielectric material, where ε is the dielectric constant of the composite and E is the breakdown strength.

$$J = \frac{\varepsilon_o \varepsilon E^2}{2} \quad (2)$$

EXPERIMENTAL

Materials:

The matrix used in this study was a commercial epoxy/polyamide system comprised of Epon 828 (Aldrich) and polyamides Ancamine 2352, Ancamine 2205, and Ancamine K-54 (Air Products). An antifoaming agent BYK052 (Byk-Chemie USA, Wallingford, CT) was utilized to facilitate air release from the epoxy. Reportedly, microscale (1-3μm) barium titanate TAM Ticon NEB (Ferro Corporation, Niagra Falls, NY, previouslyTAM Ceramics) particles were used as filler. The $BaTiO_3$ was dispersed via ball milling with 0.635cm zirconium oxide media (Paul N. Gardner, Pompano Beach, Fl) for 24 hours with Menhaden fish oil (Z-3, Werner G. Smith, Inc.), polyamide, antifoaming agent, and a solvent of 80/20 % volume xylene/ethanol.[10] The oil dispersant concentration was optimized at approximately 3% weight to the $BaTiO_3$ using a Hegman gage according to ASTM D 1210-96. Particles were incorporated in a range of low volume concentration filler loadings ($0<\varphi<25$) and composites were synthesized according to ASTM D 2149-97 in films (0.03-0.35 mm thickness). The formulation was stirred with a bladed mixer followed by sonication in order to aid in the release of bubbles. The films were cast using a draw down bar (doctor blade) and cured at 25°C for 48 hours followed by a gradual increase until the temperature of the polymer glass transition temperature (Tg ≈95°C) was reached where temperature was then held constant for 3 hours. The films were stored in an oven at 30°C until testing in order to prevent ambient water absorption.

Measurements:

X-ray diffraction patterns were obtained to characterize the barium titanate powder using Cu Kα on a Scintag PadV. Crystal structure and size were determined based upon peak splitting and line broadening.

Scanning electron microscopy was used to visualize the powder dispersion, agglomeration, and particle size in the epoxy matrix. Images were obtained on the surface and cross-sectional fracture of cured films with a Hitachi S-4700 FE-SEM.

Dielectric testing was performed on the films and pellets synthesized according to ASTM D 2149-97 after silver electrodes were sputtered for 300 sec at 60 mA through a shadow mask. A micrometer-electroder system was used whereby 2 flat plates came in full contact with the electrodes and film, preventing warping as was observed with a more traditional point probe system. Capacitance and loss values were measured at room temperature using a HP 4192-A impedance analyzer at a frequency of 1kHz to 10MHz.

Breakdown strength measurements were performed using a Hewlett Packard 4140 B pA meter under DC voltage with a Trek Model 20/20A amplifier. Neat epoxy/polyamide was dip-coated onto copper plates creating a film (0.015-.040mm) which was then cured for 48hours at room temperature followed by 48 hours at 70^0C. Free standing films (.05-0.2mm) containing ball milled $BaTiO_3$ were also created. Silver electrodes of 0.7 cm diameter were sputtered on the flat test specimens at 60mA for 300 sec. The films were then placed in FC- 40 Fluorinert electronic liquid (3M), and a point probe-flat plate set-up was used for testing. The voltage was increased in a step-wise fashion until breakdown occurred and was observed by a sharp rise in current and a consequential small pin-hole created in the test specimen.

RESULTS

Characterization of barium titanate:

In order to understand the properties of the ceramic-polymer composite it was necessary to examine both the barium titanate and commercial epoxy/polyamide system used in synthesis. First, the properties of the barium titanate particles were determined, including crystal structure, particle size, and dispersion. From x-ray diffraction patterns of the barium titanate powder (see Figure 1), we concluded that the TAM micrometer particles were tetragonal due to the peak positions and characteristic splitting around 45 or 75 degrees 2θ.[8] According to information provided by Ferro, the particle size of the barium titanate was 1-3μm, and average crystal size was measured by XRD to be 57.7 nm, which was verified through field emission scanning electron microscopy images.

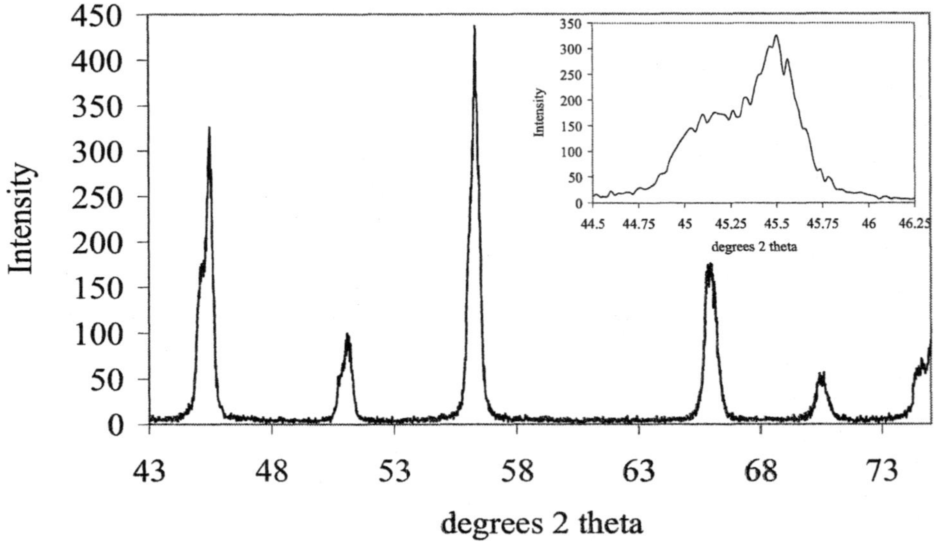

Figure 1. X-ray diffraction pattern of $BaTiO_3$ (μm). Insert shows the peak splitting indicative of a tetragonal crystal structure.

Dispersion of Barium Titanate:

The dispersion of particles has been reported to be an important factor in optimizing the electrical performance of polymer-ceramic composites.[11,12] The stability of a suspension and a resulting homogeneous composite depends upon the surfactant's ability to sufficiently separate particles through repulsive forces, which must overcome van der Waals attraction. The repulsive forces may originate from steric or electrostatic mechanisms. A common commercial dispersant, Menhaden fish oil, was used in the synthesis of films and optimized at a 3% weight to the $BaTiO_3$. The filler's dispersion was tested according to ASTM D 1210-96 with a Hegman gage before the films were cast and cured.

To more fully characterize the nature of the barium titanate, FE-SEM was used to determine particle agglomeration, size, and dispersion in the epoxy matrix. Specimens were examined on the surface and in the fractured plane. Particles in the film did not appear to be agglomerated, and were approximately 1 μm. The barium titanate particles were closely examined and crystal size confirmed to be 57.7 nm as was determined by XRD. The $BaTiO_3$ was fully dispersed throughout the epoxy matrix for the pigment volume concentration of 5 and 25 (see Figure 2). Test specimens having a $BaTiO_3$ loading of 7 and 15% volume had slight settling in the composites, possibly due to a low viscosity of the uncured composites.

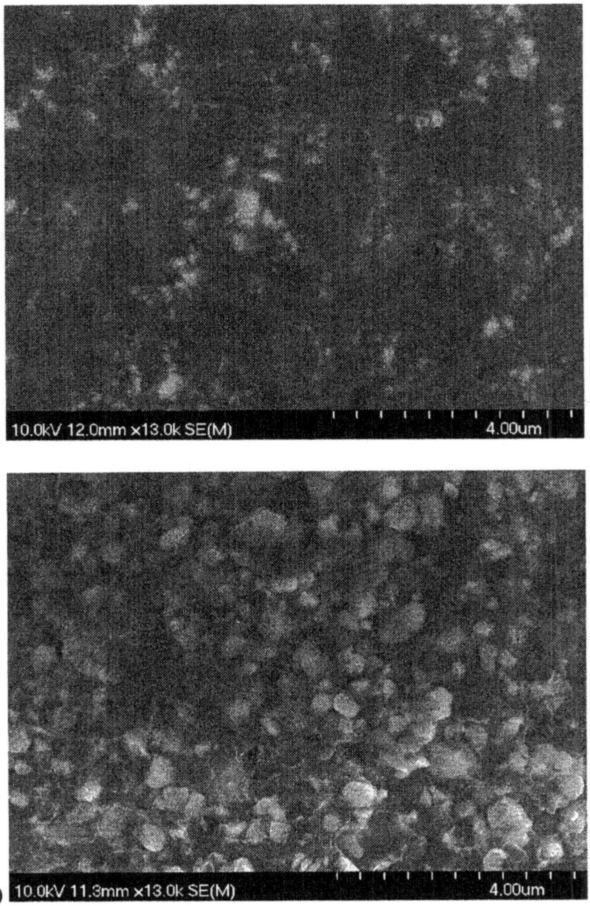

Figure 2. FE-SEM images of the dispersion of $BaTiO_3$ at a 25% volume loading in epoxy (a) on the surface and (b) in the fractured plane.

Electrical Measurements:

There are two critical components in the calculation and optimization of energy density: the dielectric constant κ and the breakdown strength E of the material in volts (Equation 2). In order to create a composite with high energy density, a trade-off frequently exists between the permittivity and the breakdown strength of the composite. A polymer, such as epoxy typically has a κ of 3-5, and a breakdown strength of 5000 kV/cm.[13] When loaded with the ceramic filler barium titanate, which has a much higher permittivity, the dielectric constant of the overall composite would be expected to increase; however, a decrease may be commonly observed in the breakdown strength due to surface interactions and/or differences in permittivity between the matrix and filler that causes local field enhancements. Therefore, it was important to determine how much the dielectric constant and breakdown strength of the composite were affected by a change in percent volume filler loading with barium titanate in order to create the highest energy density hybrid material.

First, the capacitance and loss were measured between 1kHz and 10MHz for neat epoxy and composites containing barium titanate. Samples were synthesized in a range of filler concentrations from 0 to 25%. The dielectric constants were calculated for 3 specimens at each concentration and averaged to form the trend seen in Figure 3. As the loading level of barium titanate increased, the dielectric constant of the composite material increased up to a κ of 13.27 at a pigment volume concentration of 25%. The dielectric loss remained low, at approximately 0.02 over the ceramic concentration range. All samples were dried extensively in order to volatilize any residual solvent; ethanol is polarizable with a $\kappa = 24.3$, and can produce an artificially high dielectric constant and loss for the composite if not completely removed. Error bars at each concentration level indicate the standard deviation for the test specimens and did not exceed 0.6 for the entire range of filler loadings.

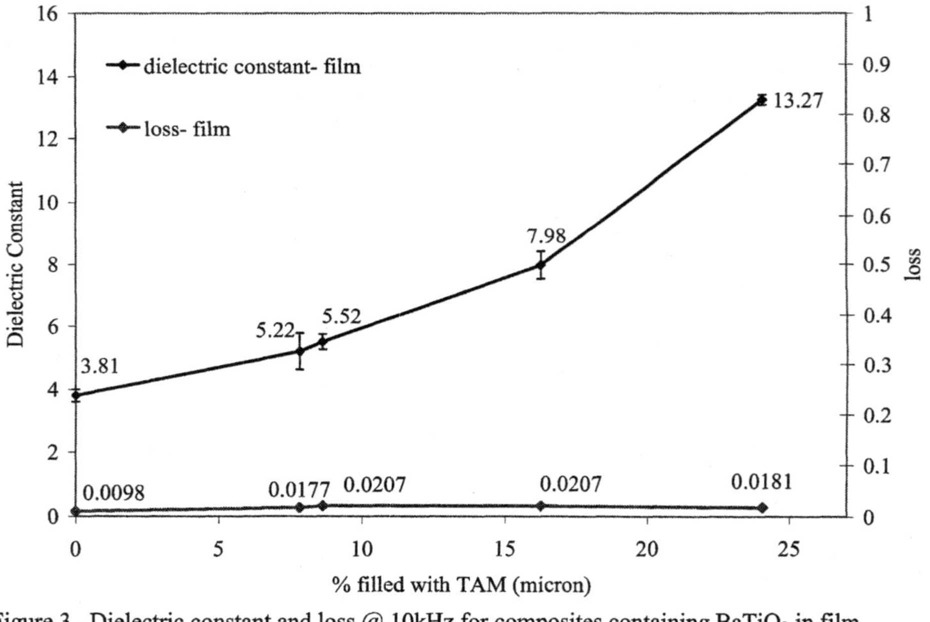

Figure 3. Dielectric constant and loss @ 10kHz for composites containing $BaTiO_3$ in film geometry according to ASTM D 2149-97.

Dielectric Breakdown Strength Measurements:

The definition of electrical breakdown strength is the electric-field intensity required to cause a transition in a material from an insulator to conductor. This transition is accompanied by a sharp increase in current. Electrical breakdown is not a simple phenomenon; rather, it is highly complex and driven by the geometry of the sample and electrodes, the material chemical composition(s), and the history of the sample including moisture, temperature, electrical, and environmental. Table I indicates the breakdown strength of neat epoxy/polyamide, which had been applied by dip-coating of copper sheet.

Table I. Breakdown strength and the data scatter for composites containing epoxy/polyamide and barium titanate as a function of the ceramic volume concentration.

Sample	Ceramic Volume Concentration	Average Breakdown Strength (kV/cm)	Standard Deviation (kV/cm)	Number of Samples
Neat epoxy	0	2329	547	8
$BaTiO_3$ - 7	7	1243	296	10
$BaTiO_3$ - 9	9	1025	195	11
$BaTiO_3$ - 16	16	936	288	10
$BaTiO_3$ - 25	25	763	241	9

Neat epoxy had a breakdown strength of 2329 kV/cm, which is low compared to previously reported literature values of 5000 kV/cm, but is likely due to the use of different curing agents in the formulation.[13] One major source of error may lie in the geometry of the flat sample and sputtered electrode. Field enhancement effects are known to exist at the electrode edge and produce premature breakdown when flat samples are used instead of dimpled. For example, as can be observed in Figure 4, the dielectric strength test for the neat epoxy most often failed at the electrode edge.

Figure 4. Panel of neat epoxy after dielectric strength testing.

It was not only critical to determine the permittivity behavior of the epoxy/barium titanate composite, but also the dielectric strength of the composite as influenced by the concentration of the ceramic in order to understand the role of the filler in the energy density. Therefore, the breakdown strength of composite films containing ball-milled $BaTiO_3$ particles was plotted verses the percent volume concentration (Table I and Figure 5). As the concentration of the barium titanate increased the breakdown strength of the composite sharply decreased. The decreasing trend was expected as the amount of surface interactions increases between the two vastly different dielectric materials, epoxy and $BaTiO_3$. Dielectric breakdown would be expected to occur at the interface of the materials unless a pore or other defect is present in the film. In addition to the presence of possible pores or defects, the wide range of data at each concentration may also be due to similar field enhancement at the electrode edges as was observed in the neet epoxy samples though large deviations are commonly observed in dielectric strength measurements.

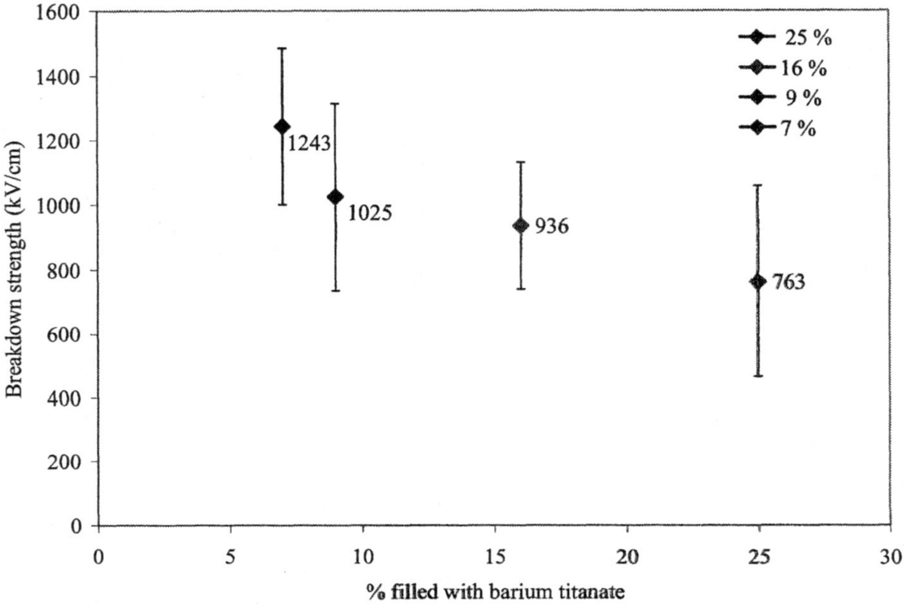

Figure 5. Breakdown strength of epoxy/polyamide composites filled with $BaTiO_3$.

Energy Density Calculations:

The energy density, calculated from the combined dielectric constant and dielectric strength (Equation 2), is a valuable assessment of the overall electrical properties of a material. For use as capacitor, a composite should have a high energy density, which may be achieved through optimization of κ or dielectric breakdown strength.

Since the filler volume concentrations for the prepared samples are low, we assumed the energy density will be governed by a linear response characteristic of epoxy and other polymer systems. Notable is that theoretical energy density is linearly driven by the dielectric constant of the composite ε but a non-linear square function of dielectric strength E. Therefore, we anticipated the energy density would be highest for those samples producing the highest breakdown strength, i.e., in our case, samples with the lowest volume filler concentrations. Using the data determined previously experimental testing, the energy densities were calculated to be approximately 0.31 J/cm^3 over the entire concentration range.

Table II. Calculated energy densities of composite $BaTiO_3$ incorporated in epoxy polymer.

Sample – Ceramic Conc.	Pigment Volume Concentration	Avg. Dielectric Constant	Breakdown Strength (kV/cm)	Energy Density (J/cm^3)
Neat epoxy	0	3.8	2329	0.912
$BaTiO_3$ - 7	7.83	5.22	1243	0.357
$BaTiO_3$ - 9	8.64	5.52	1025	0.257
$BaTiO_3$ - 16	16.26	7.98	936	0.310
$BaTiO_3$ - 25	25.06	13.27	763	0.342

CONCLUSIONS

In was desired to characterize and determine the energy density of a composite of epoxy and barium titanate. First, the polycrystalline $BaTiO_3$ powder was determined to be tetragonal, where XRD and SEM confirmed the particle size to be 1-3 μm in size with a crystal size of 57.7 nm. Next, Menhaden fish oil was used to incorporate $BaTiO_3$ particles and SEM images confirmed good dispersion of the composite films for all ceramic volume concentrations synthesized, 5 through 25 percent, however, films of pigment volume percentages 7 and 15 showed slight settling of the ceramic during synthesis. The dielectric constant was found to be approximately 13 for composites at the maximum 25% $BaTiO_3$ concentration while loss values remained low at approximately 0.02 for the entire range of incorporation. The breakdown strength varied from 1243 to 763 kV/cm for barium titanate incorporated samples. Finally, energy densities of 0.31 J/cm^3 were calculated based on a linear dielectric and were consistent in the low filler volume concentrations.

AKNOWLEDGEMENTS

Technical assistance by Clarissa Wisner and Manesh Nadupparambil with SEM and XRD measurements, respectively, was appreciated. Ahmed Shahin photographed the dielectric strength sample. The authors gratefully acknowledge project support from the NSF, Penn State University, University of Missouri-Rolla, Center for Dielectric Studies, and the Department of Education GAANN program, grant number P200A040065 .

REFERENCES

[1]Nancy E. Frost, Paul B. McGrath, and Clayton W. Burns, "Effect of Fillers on the Dielectric Properties of Polymers," *1996 IEEE International Symposium on Electrical Insulation, Montreal, Quebec, Canada, June 16-19, 1996*, 300-303.

[2]Tia S. Troutman, Swapan Bhattacharya, Rao Tummala, and C.P.Wong, "Development of Low Viscosity, High Dielectric Constant (K) Polymers for Integral Passive Applications," *1999 International Symposium on Advanced Packaging Materials*, 169-173.

[3]Yang Rao, S.Ogitani, Paul Kohl, and C.P.Wong, "High Dielectric Constant Polymer-Ceramic Composite for Embedded Capacitor Application." *2000 International Symposium on Advanced Packaging Materials,* 32-37.

[4]Yang Rao, S.Ogitani, Paul Kohl, and C.P.Wong, "Novel High Dielectric Constant Nano-Structure Polymer-Ceramic Composite for Embedded Capacitor Application," *2000 Electronic Components and Technology Conference,* 183-187.

[5]Lianhua Fan, Yang Rao, C.Tison, K.S. Moon, S.V. Pothukuchi, and C.P. Wong, "Processability and Performance Enhancement of High K Polymer-Ceramic Nano-Composites," *8th International Symposium on Advanced Packaging Materials*, 120-126.

[6]Yang Rao, Jireh Yue, and C.P.Wong, "High K Polymer-Ceramic Nano-Composite Development, Characterization, and Modeling for Embedded Capacitor RF Application," *2001 Electronic Components and Technology Conference.*

[7]Satoshi Ogitani, Sue Ann Bidstrup-Allen, and Paul Kohl, "An Investigation of Fundamental Factors Influencing the Permittivity of Composite for Embedded Capacitor,"*1999 Electronic Components and Technology Conference,* 77-81.

[8]Sung-Dong Cho, and Kyung-Wook Paik, "Relationships Between Suspension Formulations and the Properties of $BaTiO_3$/ Epoxy Composite Films for Integral Capacitors," *2001 Electronic Components and Technology Conference,* 1418-1422.

[9]G.Arit, D.Hennings, and G.deWith, "Dielectric Properties of Fine-Grained Barium Titanate Ceramics," *J. Appl. Phys.,* **58,** (4) 1619-1626 (1985).

[10]Jian-Huei Feng, and Fatih Dogan, "Effects of Solvent Mixtures on Dispersion of Lanthanum-Modified Lead Zirconate Titanate Tape Casting Slurries," *J. Am. Ceram. Soc.* **83** (7), 1681-86 (2000).

[11]Lianhua Fan, Yang Rao, C.Tison, K.S. Moon, S.V. Pothukuchi, and C.P.Wong, "Use of Dispersant in High K Polymer-Ceramic Nano-Composite to Improve Manufacturability and Performance of Integral Capacitors," *2002 Electronic Components and Technology Conference,* 936-940.

[12]Ungyu, Paik, Vincent A. Hackley, Sung-Churl Choi, and Yeon-Gil Jung, "The Effect of Electrostatic Repulsive Forces on the Stability of $BaTiO_3$ Particles Suspended in Non-aqueous Media," *Colloids and Surfaces A: Physicochemical and Engineering Aspects,* **135**, 77-88 (1998).

[13]N. Yoshimura, M. Nishida, S. Fujita, and F.Noto, "Electrical Breakdown of Alumina Filled Epoxy Resin," *Proceedings of the 14th Symposium on Electrical Insulating Materials, October 5-7*, 89-92 (1981).

BARIUM STRONTIUM TITANATE GLASS CERAMICS FOR HIGH ENERGY DENSITY CAPACITORS

E.P.Gorzkowski, M.-J. Pan, B. Bender, and C. C. M. Wu
Naval Research Laboratory
Code 6351
4555 Overlook Ave., SW
Washington, DC 20375

ABSTRACT

Barium titanate was studied in 2003 as a potential ferroelectric glass-ceramic for high energy density dielectric materials. The dielectric constant of these heat-treated samples was up to 300 and the dielectric breakdown strength up to 800 kV/cm. In this study a new ferroelectric phase barium strontium titanate (BST) was introduced due to the potentially greater energy density and to the stability at higher field strengths. This paper will present the benefits and complications of this substitution, while highlighting the dielectric properties that result. In this case the dielectric constant reached 1000 for BST 80/20 and breakdown strength was as high as 800 kV/cm for BST 50/50. This did not result in exceptional energy density (~ 0.70 J/cc), but the existence of dendrites and pores means that more optimization is needed.

INTRODUCTION

Currently, warships and combat vehicles are built with a propulsion system that is separate from auxiliary systems and weapons. Therefore a large amount of power is locked into the mechanical propulsion train and is not available for any other use. To remedy this issue the currently planned all-electric ship is to have both the propulsion system and pulsed power weapons draw from the same energy source. This integration will significantly improve efficiency, effectiveness, and survivability while simultaneously increasing design flexibility, reducing total ownership costs, and enhancing the quality of service.[1] In the past several years, the US Navy has made a significant investment in the technology of power electronics for the all-electric warship. Despite the tremendous progress in the area of semiconductor switches[2], the passive components, especially capacitors, remained a limiting factor in the design of high power systems due to their low volumetric efficiency.[3] In pulsed power weapon applications where capacitor banks can occupy many cubic meters of space, the importance of developing high energy density dielectric materials is obvious.

There are several classes of materials that can be improved to develop these high-energy density capacitors, but each has problems. Ceramics generally have a high K, dielectric constant, (4000), but possess low breakdown strength (100 kV/cm). Polymers on the other hand have high breakdown strength (4000 kV/cm) but a poor dielectric constant (3). Since the energy density is related to the product of breakdown strength squared and dielectric constant, ceramics can be improved by increasing the breakdown strength and polymers can be improved by increasing K through filler additions.[4] Consequently, improvements in breakdown strength have a more pronounced effect on energy density so that ceramics systems have the best chance to become high-energy density capacitors. In fact, the low breakdown strength of ceramic capacitors is due to flaws, such as porosity, which causes (electrical) stress concentration and corona. Therefore, the energy density could easily be increased by simply eliminating flaws, i.e. pores, in the ceramic materials.[5]

After assessing existing ceramic technologies it was found that several possibilities exist to dramatically increase the energy density of dielectrics. The most attractive candidate from this study was the development of ferroelectric glass-ceramics.[6] In this approach, a glass medium that is comprised of glass network formers and ferroelectric constituents is prepared by melt-casting. Ferroelectric particles/grains are then precipitated within the glass matrix during subsequent heat treatment.[7,8] The synergistic effect of good dielectric properties from the ceramic and the defect-free nature of a glass results in a material that can have high dielectric constants and high breakdown strengths.[9] The major advantage of the approach is that by starting with a porosity-free glass matrix, one can eliminate the source of electric field concentration and increase the dielectric breakdown strength to the vicinity of its intrinsic value of a few MV/cm.[5] In this way, an energy density greater than one order of magnitude higher than that of conventional dielectrics can be achieved. Certainly the presence of a low dielectric constant (K) glass phase imposes a tradeoff in the final energy density, but its effect can be minimized by employing a high volume fraction of high K ferroelectric particles. Most importantly, the benefit of ultrahigh breakdown strength will far out-weigh the slight decrease in overall dielectric constant.

In 2003 Penn State University together with the Naval Research Laboratory (NRL), completed a feasibility study to investigate glass ceramics. The systems observed in that study utilized high dielectric crystalline phases that were ferroelectric in nature. The crystallization and dielectric behavior of the glass-ceramics in the PbO-BaO-SrO-Nb_2O_5-B_2O_3-SiO_2 and Na_2O-PbO-Nb_2O_5-SiO_2 systems were investigated. The glasses were formed by rapidly quenching the melt between two aluminum plates. After the crystallization/nucleation temperature was determined via thermal differential analysis, a series of heat treatments at different temperatures and dwell times were carried out. XRD revealed the crystallization of $Sr_{0.33}Ba_{0.67}Nb_2O_6$ phase in the PbO-BaO-SrO-Nb_2O_5-B_2O_3-SiO_2 system and $Pb_2Nb_2O_7$, $NaNbO_3$ and $PbNb_2O_6$ phases in the Na_2O-PbO-Nb_2O_5-SiO_2 system. The maximum dielectric constant, K, was approximately 250-300. The dielectric breakdown strength was in the range of 400-800 kV/cm which is 5 to 10 times higher than that of conventional ceramic dielectrics.[4] NRL, then, conducted a further study using the same methods on the BaO-TiO_2-Al_2O_3-SiO_2 system, with the intent of producing $BaTiO_3$ as the crystalline phase. This study had similar results as the joint study with Penn State in that K was approximately 350 at room temperature and the dielectric breakdown strength was in the range of 600-800 kV/cm.[10] These results indicated that a final energy density of 6 to 8 joules/cm^3 is possible, but the one issue with $BaTiO_3$ was the energy density results at high field conditions. Therefore, in this study SrO was added to the melt in order to form $Ba_{(1-x)}Sr_xTiO_3$ and hence lower the Curie point of the ferroelectric phase. In this way, at room temperature the BST is a paraelectric phase and therefore is expected to maintain its dielectric properties at high field. In addition, by adding strontium the energy density could potentially be increased as shown by other experimental and theoretical studies.[11] This increase is necessary to achieve an energy density higher than 10 J/cm^3, which is a projected long range goal for bulk materials to be used on the all electric ship.

EXPERIMENTAL PROCEDURES

Barium Strontium Titanate (BST) glass was made by mixing high purity $BaCO_3$, $SrCO_3$, TiO_2, Al_2O_3, SiO_2, and refining agents such as B_2O_3, BaF, and P_2O_5 where indicated. The appropriate amount of each powder was added to form several batches of BST with barium stoichiometries of 50, 70 and 80 mol%, while keeping the glass former content between 20 and

35 mol%. These batches were jar milled overnight in a high density polyethylene bottle with no milling media for homogenous mixing. The powders were then placed in a platinum crucible and heated to 1500 °C in a box furnace in order to decompose the carbonates and then to subsequently melt the constituent powders to a viscous liquid. The melt was allowed to homogenize at this temperature for 2 hrs. prior to pouring. The crucible was then quickly removed from the furnace and poured onto a set of rollers for quenching. Next, the transparent BST glass was annealed at 600 °C for 10 hrs to remove residual stresses. Glass from each batch was then ground using an agate mortar and pestle to a fine powder for X-ray Powder Diffraction (XRD) and DTA (Differential Thermal Analysis). The XRD was completed on a Phillips XRG 3100 diffractometer with Cu $K\alpha$ radiation at 50kV and 30 mA to verify that no crystallization had taken place. To determine the crystallization temperature DTA (TA instruments) was performed using a 10 °C/min ramp rate to 1200 °C. Remaining glass from each composition was then separated into smaller lots and annealed at temperatures of 850 – 1200 °C in order to crystallize the glass, leaving the desired glass ceramic product.

One sample from each composition and heat treatment were then ground into a powder with an agate mortar and pestle for XRD to verify the crystalline phases. Additional samples of each composition were then prepared for dielectric testing. This entailed lapping the samples using 400 and 600 grit SiC slurry to create flat parallel faces. Some samples were gold coated for capacitance measurement, while others were masked for breakdown and energy density measurements. The dielectric constant and loss were measured using an HP 4284A at 0.1, 1, 10, and 100 kHz from 150 down to -60 °C. The breakdown measurements were made using a Hipot tester (QuadTech) at 100 V/s. Energy Density at room temperature was determined using a discharge circuit with calibrated resistors, measured with an Agilent 54622A oscilloscope while integrating the resultant curve.

Slices from each of the various samples were mounted in epoxy and polished from 600 grit SiC carbide to 0.05 μm Al_2O_3. The epoxy mounts were then carbon coated and masked with conductive tape for Scanning Electron Microscopy (SEM). Images of the polished surfaces were obtained using a Leo 1550 SEM.

RESULTS AND DISCUSSION

Table I shows a comparison of compositions used in the previous NRL study versus some used in the current one. The older study used barium titanate (BT), while the new study used barium strontium titanate. (BST 50/50 is $Ba_{0.5}Sr_{0.5}TiO_3$, etc.)

Table I: Sample Compositions in mol%

	BT	BST 50/50	BST 80/20
BaO	36.7	19.34	35.01
TiO_2	32.4	31.32	35.43
SiO_2	14.2	17.14	11.32
Al_2O_3	7.7	6.43	4.24
F_2	0.6	---	1
MgO	1.3	---	---
Excess BaO	12.7	---	---
SrO	---	19.34	8.75

It is important to notice that the BST samples contained Al_2O_3 as this addition played a key role in obtaining BST as the primary crystalline phase. Figure 1 shows that without alumina addition (Figure 1 a.) the sample contained approximately 50 % BST, while the other half was a crystalline silicate phase. The work of Herczog showed that alumina acts as a network former, which in the correct ratio with silica prevents high cross-linking so that a stable random network forms.[12,13] In addition, by keeping the silica content low the glass network breaks down easily, which promotes the nucleation of the crystalline phase and keeps the starting grain size small.[14] This situation is ideal because to achieve a reasonable dielectric constant, the amount of remnant silica should be low. Figure 1 b shows that this is also the case in BST glass ceramics, as the primary phase in samples containing alumina is the desired BST phase. In addition, there is some remnant silica, and a barium rich phase that is unidentified in the spectra, but is either unreacted BaO or $BaSiO_3$.

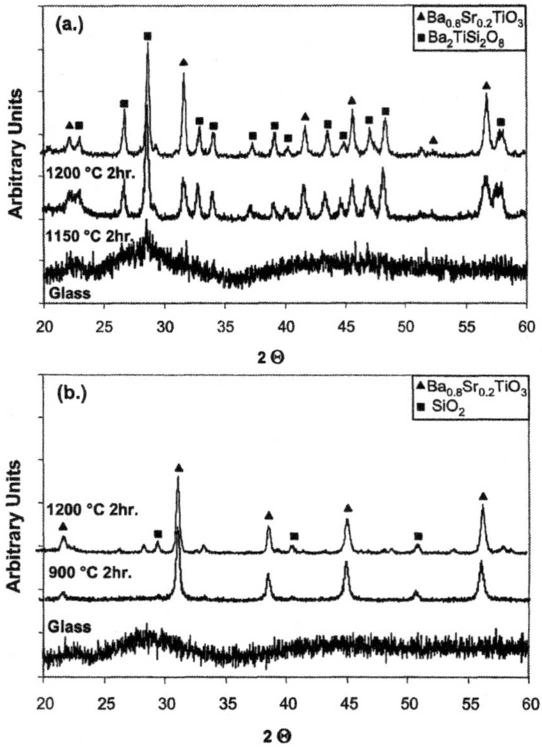

Figure 1. XRD spectra of BST 80/20 with (a.) no Al_2O_3 and (b.) a small Al_2O_3 addition.

Once samples containing a majority of crystalline phase were formed, dielectric property measurements were taken. Figure 2 shows the effect that composition has on the dielectric constant and Curie temperature. The BST 50/50 sample in figure 2a, shows that the peak dielectric constant is around 250 and occurs at -90 °C, while the BST 80/20 has a peak dielectric constant of 1000 at 80 °C. This sampling of data shows that as more SrO is added the Curie temperature and dielectric constant decreases. The trend held for all samples that were tested regardless of heat treatment. The Currie temperature behavior is consistent with the work of Hilton and Ricketts[15], which shows the same dependence with SrO content. The decrease in dielectric constant is due to the higher glass content needed to form glass precursors as $SrTiO_3$ was added.

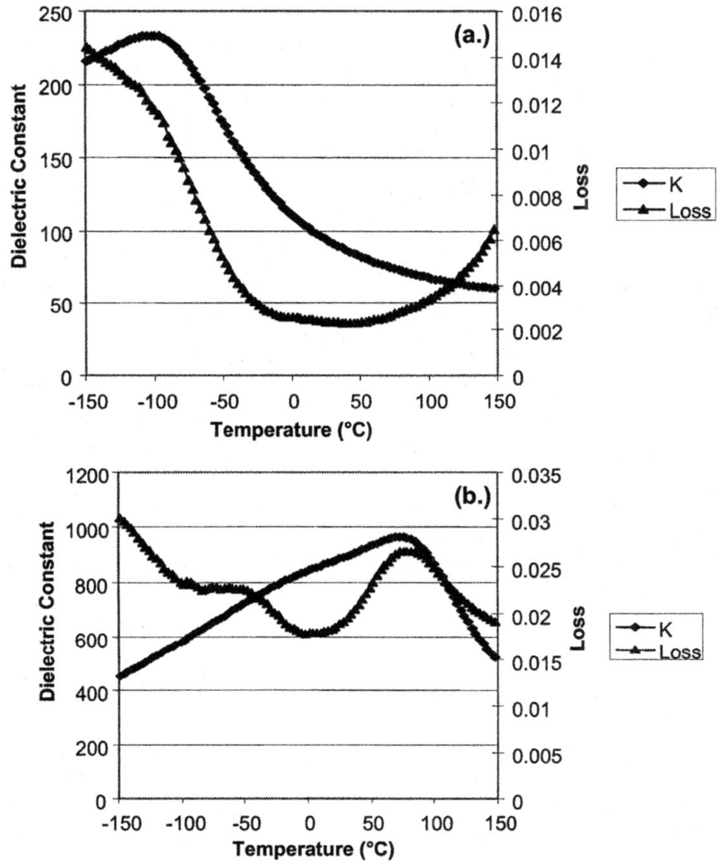

Figure 2. Dielectric constant and loss versus temperature for BST (a.) 50/50 and (b.) 80/20 annealed at 1200°C for 10 hrs at 100 Hz.

This study shows that a dielectric constant of 1000 (Fig. 2) can be reached due to a lower amount of glass former as compared to the barium titanate study where the highest dielectric constant was 350. The higher glass content reduces the dielectric constant, but may be necessary to create the glass ceramic. Therefore, a processing window exists for each composition of BST where the properties are ideal to obtain high energy densities.

The breakdown strength of these samples also varied. As the barium content increased, the breakdown decreased. BST 50/50 had a breakdown strength of 800 kV/cm, BST 70/30 was

500 kV/cm, and BST 80/20 was 300 kV/cm. The same trend occurred in the resulting energy density, where BST 50/50 had the highest storage capacity. The values ranged from 0.744 J/cc for 50/50 to 0.279 J/cc for the 80/20 sample. These values were obtained from samples that were annealed at 1200°C for 2 hrs. and were verified by measuring hysteresis loops for each sample and integrating the area between the polarization axis and the discharge curve in a polarization versus electric field plot. Though these values were disappointing, the potential to achieve much higher values still exists as these samples were not completely optimized. This can be seen by Figure 3 which shows the microstructures of BST 50/50 and 80/20 samples. Both microstructures contain dendrites which are far from the desired equiaxed particles. This type of microstructure creates a high field concentration at the tips of the dendrites causing premature failure. Along with the pores found in the BST 80/20 sample, the appearance of dendrites reduces the properties and shows the need for more optimization of processing conditions. Dendrite formation in glass ceramics systems have been reported before, and in all cases have reduced the desired properties.[16,17]

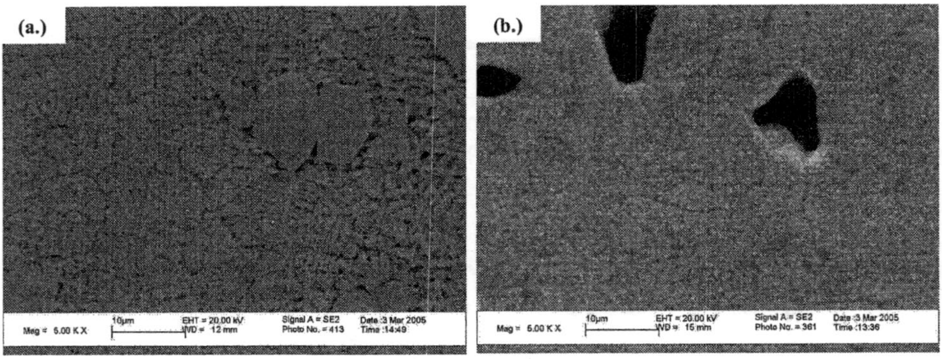

Figure 3. SEM micrographs of BST (a.) 50/50 and (b.) 80/20 showing the defects that still exist.

In order to prevent dendrites, the cooling rate can be adjusted as well as adding refining agents to prevent the super-saturation of any constituents. Additionally, other additives such as B_2O_3 could be used as a glass former instead of SiO_2 to further reduce the viscosity of the melt. This should enable the melting of the mixture at lower temperatures, and prevent deleterious phases from forming that may alter the properties. Future studies plan to observe these effects, along with continuing to characterize the properties of the BST system.

In addition, annealing treatment affected the energy density. Figure 4 shows the energy density dependence of field on samples annealed under different conditions. Interestingly, though the values are different the slope of the energy density versus field lines remain the same regardless of heat treatment, which suggests that the samples annealed at lower temperatures contain less crystallized material.

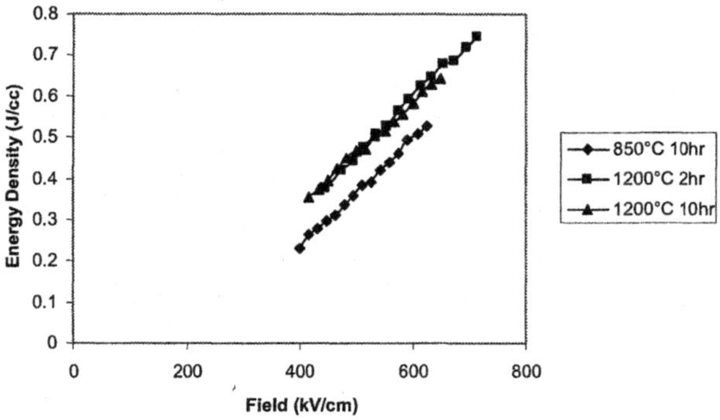

Figure 4. Energy Density versus field for several different annealing treatments in a BST 50/50 sample.

Figure 4 also shows that the annealing time can also affect the peak energy density. The sample fired at 1200 °C for 2 hours was able to withstand a higher field and thus achieve a better energy density than the sample annealed for 10 hrs. This is due to the fact that much grain growth has occurred in the 10 hr annealed sample so that breakdown occurs earlier, which suggests the ideal situation would be a high rate of nucleation and low rate of grain growth. Future studies plan to observe this effect to further optimize the microstructure in order to improve the energy density.

CONCLUSION

Previous studies on barium titanate suggested that glass-ceramic technology has the potential for high energy density dielectric materials. In this study, the material was changed to barium strontium titanate due to the potentially greater energy density and to the stability at higher field strengths. It was shown that the presence of Al_2O_3 was vital to obtaining crystallization of the desired ferroelectric phase. Also, this study showed that a dielectric constant of 1000 for BST 80/20 and breakdown strength of 800 kV/cm for BST 50/50 could be obtained. This did not result in exceptional energy density (~ 0.70 J/cc), but the existence of dendrites and pores means that more optimization is needed, and higher energy density is achievable.

ACKNOWLEDGMENTS

This research was performed while the presenting author held a National Research Council Research Associateship Award at the U.S. Naval Research Laboratory. The authors would also like to acknowledge funding from the Office of Naval Research under contract # N0001404WX20802.

[1.] R. O'Rourke, "Electric Drive Propulsion for U.S. Navy Ships: Background and Issues for Congress, *CRS Report # RL30622*, July 31, 2000.

[2.] B. J. Baliga, "The Future of Power Semiconductor Device Technology," *Proceeding of the IEEE*, **89** [6] 822-832 (2001).
[3.] W.J. Sarjeant, I.W. Clelland, and R.A. Price, "Capacitive Components for Power Electronics," *Proceeding of the IEEE*, **89** [6] 846-855 (2001).
[4.] M.-J. Pan, R.J. Rayne, B.A. Bender, and M. Lanagan, "Development of Advanced Dielectrics for High Energy Density Capacitors," in the Proceedings of the 2004 ASNE Electric Machines Technology Symposium, January 27-29, 2004, Philadelphia, PA
[5.] R. Gerson and T.C. Marshall, "Dielectric Breakdown of Porous Ceramics," *J. Appl. Phys.*, **30** [11] 1650-1653 (1959).
[6.] A. Herczog, "Application of Glass-Ceramics for Electronic Components and Circuits," *IEEE Transactions on Parts, Hybrids, and Packaging,* **9**[4] 247-256 (1973).
[7.] P.W. McMillian, Glass Ceramics, Academic Press, London and New York, 1964.
[8.] S.D. Stookey, "Catalyzed Crystallization of in Theory and Practice," *Ind. Eng. Chem.*, **51**[7] 805-808 (1959).
[9.] S. L. Swartz, *Dielectric Properties of Strontium Titanate Glass-Ceramics*, Ph.D. Thesis, The Pennsylvania State University, 1985.
[10.] M.-J. Pan, M. Lanagan, B. A. Bender, and C.-T. Cheng, "High Energy Density Ferroelectric Glass-Ceramics," pp. 187-194 in Ceramic Transactions, Vol. 169, *Synthesis, Properties, and Crystal Chemistry of Perovskite-Based Material,* edited by A. Goyal, American Ceramic Society, Westerville, OH, 2005.
[11.] N.H. Fletcher, A.D. Hilton, and B.W. Ricketts, "Optimization of Energy Storage Density in Ceramic Capacitors," *J. Phys. D: Appl. Phys.,* **29**[1] 253-258 (1996).
[12.] A. Herczog and S.D. Stookey, French Patent, 1272036 (1961).
[13.] A. Herczog, " Microcrystalline $BaTiO_3$ by Crystallization from Glass," *J. Am. Ceram. Soc.,* **47**[3] 107-115 (1964).
[14.] M.M. Layton and A. Herczog, "Structure and Crystallisation of Glasses of Low Network-Former Content," *Glass Technology*, **10** [2] 50-53 (1969).
[15.] A.D. Hilton and B.W. Ricketts, "Dielectric Properties of $Ba_{1-x}Sr_xTiO_3$," *J. Phys. D: Appl. Phys.,* **29**[5] 1321-1225 (1996).
[16.] J-J Shyu and J-R. Wang, "Crystallization and Dielectric Properties of $SrO-BaO-Nb_2O_5-SiO_2$ Tungsten-Bronze Glass-Ceramics," *J. Am. Ceram. Soc.,* **83**[12] 3135-3140 (2000).
[17.] Q. A. Juma'a and J.M. Parker, "Crystal Growth in Fluoride Opal Glasses," pp. 218-236 in Advances in Ceramics, Vol. 4, *Nucleation and Crystallization in Glasses,* edited by J.H. Simmons, D.R. Uhlmann, and G.H. Beall, American Ceramic Society, Westerville, OH, 1982.

IMPROVED ELECTRONICS RELIABILITY USING THIN FILM SMART MATERIALS FOR MITIGATING HARSH VIBRATIONAL ENVIRONMENT

W.D. Nothwang, M.W. Cole, J.D. Demaree, J.K. Hirvonen, S.G. Hirsch, C. Hubbard, E. Ngo
Army Research Laboratory
BLDG 4600, WMRD, MD, MMB
APG, MD 21005-5069

ABSTRACT

Microelectromechanical (MEMS) guidance systems currently suffer from severe inaccuracies caused by a critical failure of a MEMS inertial measurement unit (IMU) i.e., the angular rate sensor. This IMU failure is caused by its susceptibility to a harsh extrinsic vibrational environment. This environment, generated from launch, high-G, and/or in-flight vibration forces, causes an out-of-plane motion and/or a false angular rate signal to be generated. This effort will demonstrate a novel enabling technology-materials solution, which will allow guided projectiles to achieve required performance specifications in harsh mechanical vibration environments. We will demonstrate the feasibility of employing active/smart materials for passively damping spurious vibrations of device/die level structures that reduce the accuracy of MEMS inertial guidance systems. This effort will include the design of these efficient energy absorbing materials in thin film form for the first time at the micro-scale level for this purpose with a goal of obtaining > 50% vibration reduction at adverse vibrational frequencies using such composite films. In this work we present a novel approach commensurate with MEMS scale devices to mitigate the effect of spurious vibrations, and which relies on the development of a heterostructure composite bilayer design configuration composed of piezoelectric and shape memory alloy thin films, and the inherent non-linear deformations produced when these materials are mechanically loaded. Validation and verification of this design concept is achieved via continuum mechanics modeling.

INTRODUCTION

For many electronic applications there is a significant amount of energy in the contiguous environment. These sources of energy include thermal energy, solar radiation, electromagnetic, vibration and chemical reaction potential. The electronic devices need to perform in a very predictable manner, when exposed to any of these energy environments. Many of these energy sources can have very drastic effects on unprotected systems. Hermetic packaging is one mechanism of minimizing the environmental effects on the electronic device, but packaging does little to mitigate the effects of environmental vibration. Environmental vibration isolation is of particular importance in MEMS angular rate sensors (ARS) and gyroscopes that measure subtle changes in an internal vibrating mass. External vibrations can easily couple with these internal vibrations and cause serious error in measurement.

To maximize the amount of energy damped, a material must have a high elastic modulus and a high material dispersion coefficient ($\tan\delta$), as shown in Figure 1. Traditional vibration damping materials, e.g., viscoelastics, have a very high material dispersion coefficient, tan δ. Unfortunately, viscoelastics have a very low material stiffness, which limits the amount of energy that can be damped. Material stiffness is a function of temperature for most materials, but in viscoelastics the dependency is extreme, limiting their viability to a very narrow temperature range. The ideal damping materials would be relatively independent of temperature, have a high stiffness and tan δ, and the materials need to be immune to harsh environments. Three classes of materials that do meet these demands are the active/smart materials, namely magnetostrictive, piezoelectric, and shaped memory alloys (SMAs). These materials offer enhanced material dispersion, relatively high stiffness and the ability to tune and optimize the performance over the desired environmental window. [2-5]

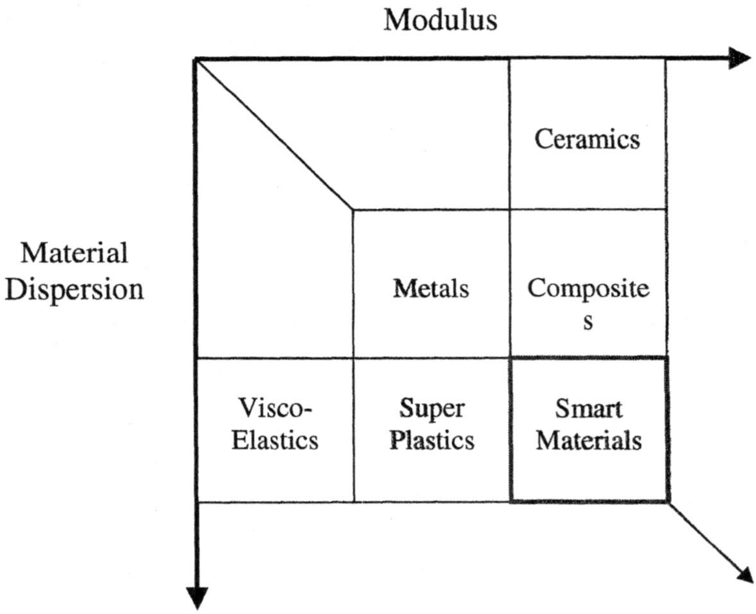

Figure 1. To achieve the optimum damping within a material, it must possess both a high elastic modulus and a high material dispersion.

A variety of authors have noted that the damping attributes of each of these active materials is distinct and dependant on the incident energy.[1-5] Specifically, shaped memory alloys can damp the most relative energy, but they require an incident energywave of amplitude of at least 100 MPa. Piezoelectrics, on the other hand, require only 30 MPa of incident energy amplitude, but are not as efficient as SMAs at dispersing the incident energy. Other active materials, such as magnetostrictive materials, also serve to damp ambient vibrations, however, these materials are very inefficient at energy dispersion, and can only disperse vibrations less than 30 MPa. Therefore, the properties of both piezoelectric and SMAs can be exploited if utilized in a bilayer thin film composite structure, i.e., a bilayer film stack as shown in Figure 1. The technical concept for using the piezoelectric and shape memory alloy in the bilayer thin film composite structure relies on the inherent non-linear deformations produced when these materials are mechanically loaded. This fundamentally novel approach, using active thin film materials, does not rely on power supplies or a shunting circuit. Rather, the technical concept for this approach is based on the reversible processes inherent in the materials, such as piezoelectric domain wall and phase/twin boundary motion (SMA), to naturally absorb the energy when the material is mechanically loaded (Figure 2 and Figure 3).[4-7] Unlike viscoelastic materials, this approach is frequency independent (though stress and amplitude dependent [1]) and is well suited to dampen vibrations across the large frequency range required in munitions based systems. More importantly, the active thin film materials have stiffness values orders of magnitude larger than viscoelastic materials and are stable over broad temperature and moisture regimes.[1] Therefore, the use of nonlinear, active thin film represents an ideal opportunity to passively dampen undesirable vibrations associated with the MEMS-based components exposed to harsh vibration environments. Based on the magnitude of stress required to induce non-linearity necessary for vibration damping, piezoelectric $Ba_{0.80}Sr_{0.20}TiO_3$ (BST) and NiTi shape memory alloy (SMA) were the active thin film materials selected for the experimental study.

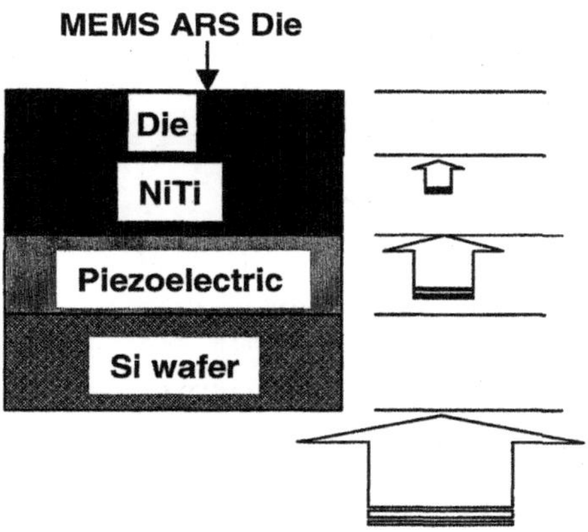

Figure 2. Schematic showing the technical concept for vibration damping using active thin films, namely piezoelectric $Ba_{0.80}Sr_{0.20}TiO_3$ (BST) and NiTi shape memory alloy (SMA) as the incident energy transits the material.

To mitigate the effects of these external vibrations on a MEMS ARS a vibration damping pedestal (VDP) will be inserted within the package between the ARS die and the base of the package. The pedestal is a multilayered thin film composite with piezoelectric, shape memory alloy (SMA), metal and ceramic layers. The goal of this study is to determine the amount of incident energy transmitted through a composite thin film structure. A minimum of 50% reduction of incident energy is necessary to allow the device to operate effectively.

To investigate this problem, a multilayered modeling approach was used. In this model, an incident compressional energy wave, I_0, enters the bottom of the pedestal from the package. I_0 is normalized to be 1. At each interface within the composite, a certain amount of the incident energy, I_j, is reflected, R_j. Additionally, another packet of energy is viscously damped, and a final amount of incident energy is removed via piezoelectric (shape memory) loss mechanisms. The piezoelectric and viscous effects have a normal and a shear component. The net shear loss, S_j, is the sum of all of the viscous and all of the piezoelectric shear components. The net normal losses, D_j, is the sum of the viscous and piezoelectric normal components. The incident energy is assumed to be parallel to the normal of the pedestal; hence, all shear components are within the plane of the pedestal. The final energy wave of interest, I_F, is the sum of the incident energy wave, I_0, and the reflected, shear, and normal losses within each layer. These waves are represented graphically in Figure 4. The general equation for the wave incident to each layer, I_j, is given in Equation 1, and the net incident wave is given in Equation 2. When the values of S_j, R_j, and D_j are significantly less than 1, it is safe to assume that each value can be calculated relative to the incident wave, I_0, not the wave incident upon each layer, I_j. This is not rigorously accurate, but as a first approximation, it is reasonable.

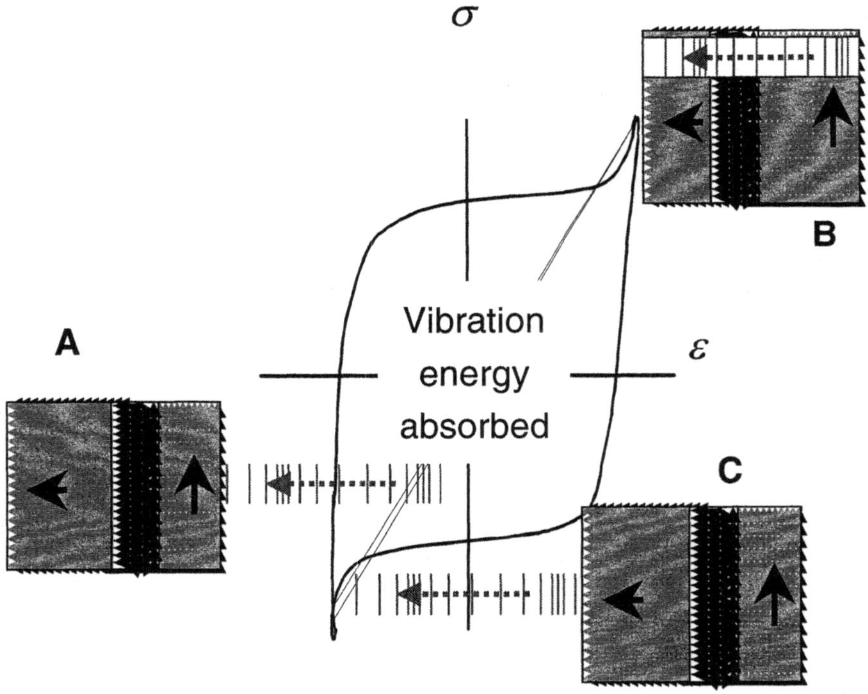

Figure 3. The hysteresis of both the piezoelectric material and the shape memory alloy accounts for a significant portion of the dispersion within the materials. A material beginning at (**A**) and cycling clockwise through (**B**) to (**C**) as a wave travels through the material will absorb an amount of energy as represented by the area within the hysteresis loop.

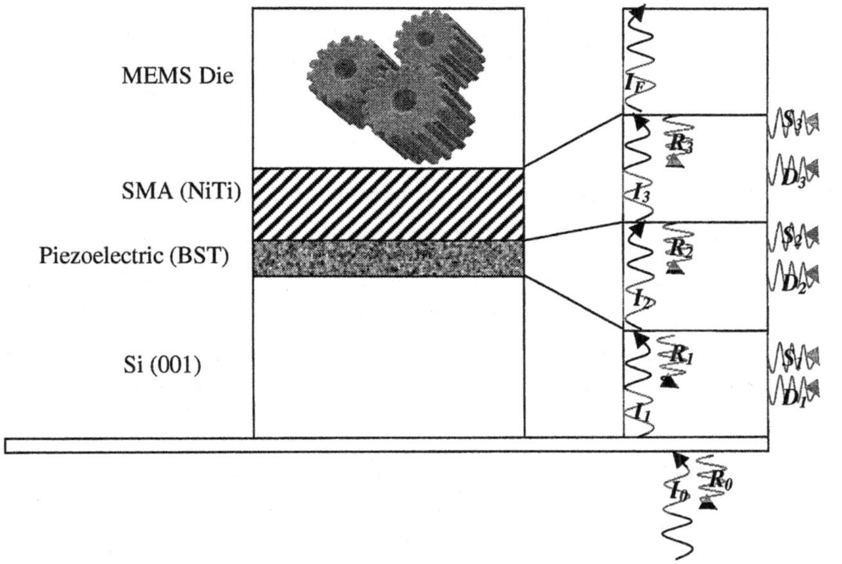

Figure 4: A visual representation is shown for the generalized solution to the wave equation for this geometry with incident and reflected waves, as well as capturing the dissipation due to shear and normal components.

$$I_j = I_{j-1} - R_{j-1} - S_{j-1} - D_{j-1} \quad \text{Eq. 1.}$$

$$I_j = I_0 - \sum_{j=1}^{j} R_{j-1} + \sum_{j=1}^{j} S_j + \sum_{j=1}^{j} D_j \quad \text{Eq. 2.}$$

MODELING

The system boundary conditions fall out directly from Figure 2. I_j is the incident wave to layer *j*; R_j is the reflected wave at the interface between layer *k* and *k+1*; S_j is the shear dissipation of the wave within layer *j*; and, D_j is the normal dissipation (damping) of the wave within layer *j*. The normal dissipation is a sum of the mechanical, the piezoelectric and the shape memory losses incurred during operation in the 33 (normal) direction. The shear dissipation is a sum of the mechanical, the piezoelectric and the shape memory losses incurred during operation in the 31 (shear) directions. The target is to have a final I_F/I_0 that is 0.5 or less indicating that more than 50% of the incident wave has been damped by the time it reaches the device. The incident wave is assumed to be primarily a compression wave traveling in the 33 direction acting only on the bottom surface of the silicon pedestal (layer 1) with a driving frequency between 3 and 20 kHz.

It is generally a reasonable assumption that all of the incident and traveling waves (ψ) are of the form *Fexp(-iωZ)*, which will satisfy the general wave equation:

$$c^2\nabla^2\psi = \left(\frac{\partial^2\psi}{\partial t^2}\right). \qquad \text{Eq. 3}$$

Specifically, for a forced oscillator with the inclusions for internal damping, Equation 3 becomes, where M, F_0 and R are constants:

$$M\left(\frac{d^2x}{dt^2}\right) + R\left(\frac{dx}{dt}\right) + \frac{x}{c} = F_0 e^{-i\omega t} \qquad \text{Eq. 4}$$

This equation can be significantly simplified to describe our situation. It is possible to imagine a 1-D wave traveling only in the longitudinal (thickness) direction of the film with minimal interactions in the transverse directions. For the damping pedestal described in Figures 2 and 4, this is a reasonable assumption, as our materials have a thickness that is significantly less than the width. Beginning with Equation 1, it is possible to solve for the 1-D wave equation as shown in Equation 5a, where the phase velocity is given by υ.

$$\nabla^2\psi - \frac{1}{c^2}\left(\frac{\partial^2\psi}{\partial t^2}\right) = 0 \qquad \text{Eq. 5a}$$

$$\upsilon = c = \sqrt{\frac{\tau}{\rho}} \qquad \text{Eq. 5b}$$

A general solution is assumed, where ω represents the angular frequency:

$$\frac{\partial^2\psi}{\partial x^2} + \frac{\omega^2}{\upsilon^2}\psi = 0 \qquad \text{Eq. 6}$$

Separating the variables and solving yields Equation 7, where the material wave number is k_i, and the individual solution is Ψ(x,t), and λ is the wavelength:

$$\Psi(x,t) = \exp\left[-i\left(\frac{\omega}{\upsilon}\right)(x - \upsilon t)\right]$$

$$\Psi(x,t) = \exp(i(\omega t - kx)) \qquad \text{Eq. 7}$$

$$\text{where } k \equiv \frac{\omega^2}{\upsilon^2} = \frac{2\pi}{\lambda}$$

Equation 7 can be modified to include the effects of damping (*-β*), as seen in Equation 8:

$$\Psi(x,t) = A\exp(-\beta x)\exp(i(\omega t - kx)) \qquad \text{Eq. 8}$$

The reflection (R) and transmission (T) coefficients are defined using the material wave numbers (k_i):

$$T = 1 - R = \frac{4k_1k_2}{(k_1 + k_2)^2}$$
$$R = \left(\frac{k_1 - k_2}{k_1 + k_2}\right)^2 \qquad \text{Eq. 9}$$

However, it is important to note, that damping will occur only if the frequency exceeds the critical frequency as defined by the critical damping frequency (ω):

$$\omega > 2\sqrt{\frac{\tau}{md}} \text{ for a infinite string}$$
$$\omega > 2\sqrt{\frac{\tau}{\rho d^2}} \text{ for a finite material} \qquad \text{Eq. 10}$$

If the frequency is greater than the critical frequency, as stated above in Equation 10, then from Equation 8, the relationship for damping is apparent, and the amount damped is a function of the materials dispersion coefficient (β), the density (ρ) and the thickness (d)..

$$\omega = 2\sqrt{\frac{\tau}{\rho d^2}} \cosh\left(\frac{\beta d}{2}\right) \qquad \text{Eq. 11}$$

RESULTS

Continuum mechanics was employed to model the longitudinal vibration modes as a function of thin film material which composed the vibration damping pedestal (VDP), solder-adhesion/bond layers and the device ceramic package. The structure shown in Figure 4 was expanded to include up to 10 layers. Table 1 shows the materials used for the modeled heterostructure. The VDP-solder/adhesion-ceramic package structure was modeled as a multilayer material with physically distinct, but mathematically continuous, interfaces. Specifically, vibration modeling for a 1-D thin plate composite was developed whereby the VCP was specified to consist of the composite bilayer thin film heterostructure, namely a piezoelectric (BST) and a shape memory alloy (NiTi), overlying the adhesion layers to the Si support wafer. Such that the Si wafer/SiO_2/Ti/Pt, whereby the SiO_2, Ti, and Pt film layers serve to adhere the base layer active thin film to the Si support wafer and deter, act as buffer layers, to deter interdiffusion between the active thin film and Si support wafer. Figure 4 displays a diagram of this multilayer structure. In this modeling approach an incident compressional wave enters the bottom of the pedestal, and it is called I_0. In this approach, it is possible to imagine a 1-D wave traveling only in the longitudinal direction of the film with minimal interactions in the transverse directions. This is a reasonable assumption, as the thin film materials possess a thickness that is significantly less than their width, but it assumes that the shear components do not significantly contribute to the damping, which is not entirely accurate for piezoelectrics and SMAs. Beginning with the wave equation for a forced oscillator with damping (Equation 3), it is possible to solve for the 1-D wave equation (Equation 8). Once solved, determining the reflection and transmission coefficients is straightforward, and they are shown in Equation 9, where k is a function of the input energy frequency (ω), the materials density (ρ), and modulus (τ).

Table 1: The compositions of the materials used in each layer of the model are shown.

1	2	3	4	5	6	7	8	9	10
Al_2O_3	Pb-Sn	Si	SiO_2	Ti	Pt	BST	Ni-Ti	Pb-Sn	SiO_2

While Equation 8 generates a set of equations describing the behavior of a traveling wave within a material, the critical thickness for damping at these frequencies is far larger than the thicknesses of the materials within the VDP heterostructure. This results in a set of equations where only the non-linear components (i.e. active materials; piezoelectric, piezomagnetic, or shape memory) of the damped wave equation remain. Consequently, it would be expected that a structure of the size of the proposed VDP would demonstrate minimal dispersion, even though the materials do demonstrate hysteretic damping. A single pass through an SMA could be expected to demonstrate a 10% dispersion at best, and values in thin films are thought to be substantially lower than those observed in bulk materials.[1] The large net dispersion anticipated from the VDP heterostructure results from the large impedance mismatch at the interfaces between layers. This effectively forces the traveling wave to traverse the BST and SMA layers, where energy dispersion takes place, numerous times, creating an effective thickness many hundreds of times larger than the physical thickness.

Reflectivity and transmission are calculated at each interface, yielding a composite transmission model relating the input energy, I_0, to the energy transmitted to the MEMS die, I_f. By examining the transmission of the energy through the composite structure, it is possible to see the effects of each layer. The energy enters the structure, as shown in Figure 4, and travels from the Al_2O_3 ceramic package to the lead-tin solder through Si wafer, adhesion/buffer layers, and composite bilayer active thin films. The energy exiting the surface, layer 10 in Figure 5, of the structure is the "final transmitted energy". The ratio of final energy to initial energy is the damping coefficient, or percent transmitted energy, shown on the y-axis of Figure 5. The layers that the energy passes through are listed in Table 1 and are shown on the x-axis in Figure 5. When only the mechanical properties of the VDP layers are examined (i.e. reflection and transmission, etc.) and the non-linear damping components are neglected, a structure that reduces the energy by less than 0.5% at steady state is yielded. This is shown in Figure 5 as the "Reflection Alone" trace. When the "strain response" nonlinear components are added for the BST and the NiTi, over 45% of the initial input energy is damped, as shown in Figure 5 with the "With 'Smart' Behavior" trace. This damping can be increased to almost 90%; if the BST and NiTi are deposited in such a way that a partial "phase change" occurs during transmission of the vibrational wave. Figure 5 reflects this result with the "NiTi Phase Change" trace. For all three of these traces it is assumed that the dispersion only occurs within the NiTi and BST layers. The observed decrease in intensity shown in Figure 5 for the "NiTi Phase Change" and "With 'Smart' Behavior" traces is due to the decreased intensity of the waves that are reflected back into that layer as shown in Figure 4. The amount of dispersion can be drastically increased, even for thin films, by joining two materials that have drastically different impedances. Continuum mechanics modeling suggests that utilizing active thin films for the VCP is appropriate for obtaining a minimum of ~50% reduction of incident energy, which is the required damping criteria necessary to allow the MEMS ARS to operate effectively.

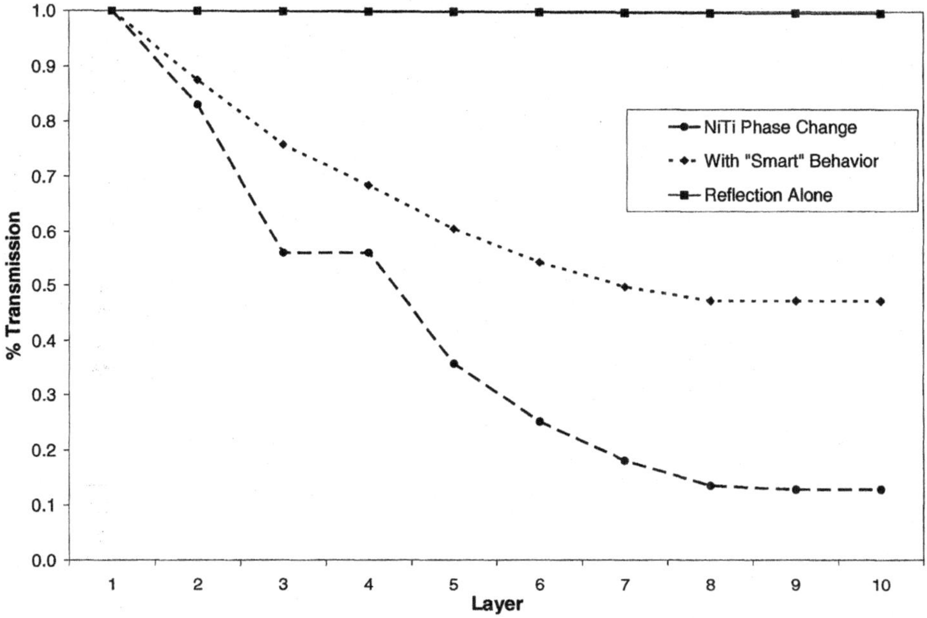

Figure 5. The effectiveness of various pedestal design is modeled as shown (a) the materials components of the VCP integrated with the MEMS die and device ceramic package, and (b) the related damping characteristics of constituent layers (i.e., solder, Si wafer, adhesion layers and VCP) within the packaged device.

CONCLUSIONS

Experimental results have demonstrated that electronic and MEMS components exposed to a harsh vibrational environment experience catastrophic failure at an alarming rate. For many applications this harsh vibrational environment is insipient, and the effect must be mitigated to achieve an optimal performance. For MEMS devices a minimum of 50% reduction of the incident waves is required to achieve an acceptable window of operation. The use of traditional vibration damping materials has proven unacceptable for a variety of reasons, and modeling suggests that the VDP heterostructure suggested here not only meets the performance metrics, but it also is able to function across the full environmental range without the need for additional packaging.

A passive damping vibration control pedestal, or VDP, based on active thin film materials was successfully designed, developed, and modeled. A continuum mechanics infinite string model demonstrated that utilizing BST and NiTi active thin films in an integrated composite bilayer design, yielded the required vibration damping response necessary for obtaining a minimum of 50% reduction of the incident energy wave. This satisfies the damping criteria for the MEMS ARS to operate effectively. The total dispersion can be increased by two methods: 1) increasing the number of layers within the structure, and/or 2) increasing the impedance mismatch between adjacent materials. Theoretical evaluation of several material integration design configurations, and their associated processing protocols, suggested the design configuration for the composite bilayer heterostructure configuration with a NiTi thin film overlying a BST base layer film on a Si substrate (with SiO_2-Ti-Pt adhesion layers). This structure would not only achieve the required dispersion of 50%, but would meet all the "foundry friendly" processing specifications necessary for mass production.

REFERENCES

1. P.M. Chaplya, G.P. McKnight, G.P. Carman, Proc. *ASME Intl. Mech. Eng. Cong. & Exhib.*, New Orleans, 1-7 (2002).
2. K.B. Hathaway, A.E., Clark, J.P. Teter., *Metallurgical and Materials Transactions A*, **26A**, 2797, (1995).
3. D.C. Lagoudas, J.J. Mayes, M.M. Khan, *Proceedings of the SPIE*, **4326**, 452 (2001).
4. P.M. Chaplya and G.P. Carman, *J. Appl. Phys.*, **90**, 5278 (2001).
5. J. Saadat, M. Salichs, Z. Noori, H. Hou, I. Davoodi, Y. Bar-On, A. Suzuki, S. Masuda, *Smart Materials and Structures*, **11**, 218 (2002).
6. M. Enokizono, T. Todaka, and M. Szpryngacz, T. Kanada, *J. Appl.. Phys.*, **91**, 7448 (2002).
7. G.P. McKnight and G.P. Carman, *Proc. Mat. Res. Soc.: Materials for Smart Systems*, **604**, 267 (2000).

ALUMINUM NITRIDE DIELECTRICS FOR HIGH ENERGY DENSITY CAPACITORS

Kevin R. Bray, Richard L.C. Wu
K Systems Corp.
1522 Marsetta Dr.
Beavercreek, OH 45432

Sandra Fries-Carr, Joseph Weimer
Air Force Research Laboratory
Wright Patterson AFB, OH 45433

ABSTRACT

Capacitors are a pervasive technology in every military and commercial application. Millions are used in electronic power systems and are considered to be a critical link to a common area of failure. Capacitors often fail under raised temperature conditions due to lowering of the voltage breakdown strength, increase in dissipation factor or problems associated with drift of the dielectric constant. Increased performance and smaller size in capacitor devices have been the main focus of our research in the development of new dielectrics. Amorphous aluminum nitride (AlN) and oxygen-doped aluminum nitride (AlON) films possess unique properties of high dielectric strength, high resistivity, low loss, high decomposition temperature, chemical inertness and good thermal conductivity. These properties make these films encouraging for use in manufacturing high energy density capacitors for pulse power applications. A dual DC pulsed magnetron sputtering technique was used to synthesize amorphous AlN and AlON films on various substrates. A statistical experimental design methodology was used to identify a process parameter set, a process trend sequence of experiments and finally a mapping of a multidimensional surface of the process parameters. The following process parameters were varied: (1) DC power, (2) pulse frequency, (3) total pressure, and (4) substrate temperature. The effects of these process parameters on the dielectric properties of the films were characterized. The dielectric constant, frequency dependence of capacitance, dissipation factor, resistivity, and breakdown strength of these films were measured using a simple parallel plate capacitor test structure.

INTRODUCTION

High Power Density Capacitors

Capacitors are key components in all forms of electrical devices. Military systems utilize millions of capacitors and consider them key components due to their susceptibility for failure. Future needs for weapons systems and aircraft performance require the development of compact, high energy density capacitors for pulsed power applications. Current pulsed power applications are dominated by polymer film capacitors. Available state-of-the-art (SOTA) dielectric materials include polymer films such as polypropylene (PP), polyester (PET) and polyvinylidene fluoride (PVdF)[1]. Polymers are used for most AC applications because of their non-polar nature and low dielectric losses. Key drawbacks to polymer films include their low operating temperature, typically ~100°C[1], and their large volume-to-weight ratio, compromising energy storage density. Most polymers also have a dielectric constant (k) in the range of 2 – 4, although k for PVdF is from 10 – 12 [1]. These low k values make it difficult to obtain the high energy density required for future military applications. Polymer breakdown voltage is typically ~550 V/μm[1]. Energy density (u_v) depends linearly on dielectric constant (k) and on the square of the breakdown strength (E_B) of a capacitor as shown in Equation 1,

$$u_v = \frac{1}{2} k\varepsilon_o E_B^2 \tag{1}$$

where ε_o is the permittivity of free space. Increasing the breakdown voltage increases the energy density more rapidly than increasing the dielectric constant. Current SOTA polymer dielectrics achieve material energy densities between 3 and 10 J/cc. Improved dielectric materials for capacitors are needed to meet the military's future power applications. Materials with a higher dielectric constant, greater dielectric breakdown strength and superior thermal stability are needed to improve capacitor performance to meet emerging needs.

Aluminum Nitride Dielectrics

Crystalline aluminum nitride (AlN) is a semiconductor with one of the largest known bandgaps (6.2 eV)[2,3] with dielectric strength between 400 and 550 V/μm[4] and thermal conductivity from 320 W/mK [3]. Amorphous AlN retains many of the crystalline properties and the insulating properties and high resistivity in conjunction with the high breakdown strength make amorphous AlN a desirable material for high density power applications. Thin AlN films have been deposited using a wide range of processes including MOCVD[5,6], RF and DC magnetron sputtering[4,7,8,9,10], and pulsed laser deposition[3,10]. Film structures from amorphous to epitaxial crystals have been obtained by varying deposition parameters and substrates[11]. Pulsed DC sputter deposition produces faster deposition rates than other deposition methods and also results in less substrate heating and thermal stressing of the films[9]. The aluminum target does not experience the same target poisoning that occurs during Al_2O_3 sputtering, making the AlN process easier to control and reproduce[8]. In this work, the material and dielectric properties of DC sputtered amorphous AlN is further investigated.

EXPERIMENTAL

Amorphous aluminum nitride films were deposited using pulsed DC magnetron sputtering. The chamber was pumped to a base pressure less than 5×10^{-6} Torr before deposition. DC power was varied from 500 – 2000 W with pulse frequencies from 25 – 250 kHz. Films were deposited using pure nitrogen and nitrogen/oxygen gas mixtures with 99.999% pure aluminum sputter targets. Gas pressures ranged from 3 mTorr to 20 mTorr. Deposition conditions strongly influence the crystallinity of the films[12], but optical and secondary electron microscopy observation confirmed our films were amorphous under all deposition conditions examined. The target-to-substrate spacing was adjusted to influence substrate heating and film uniformity with an optimal distance of 5 inches used for the majority of the runs. The target was conditioned for 15 - 30 minutes before each run using argon plasma to remove contamination from the surface. Deposition times were adjusted to achieve ~5000 Å films. Thicknesses were verified using a Dektak ^{3}ST profilometer. Films were deposited on a variety of substrates including metallized silicon, metallized glass, aluminum foil and metallized polymer. Parallel plate capacitors were constructed by evaporating 3 mm diameter dots on the top surface of the deposited films through a shadow mask.

Capacitance and dissipation factors were measured as a function of frequency using an HP 4284A LCR meter. Multiple measurements were taken at each frequency and averaged for the capacitor. Several capacitors were tested on each film to confirm uniformity across the material. The dielectric constant was calculated using the average capacitance value at 1 kHz and the measured thickness for each film. The capacitance and dissipation factor were also measured at elevated temperatures in air and under vacuum. The dielectric breakdown strength was measured using a Keithley 6517A electrometer. Breakdown voltage was determined by applying a voltage stepped in regular increments to a capacitor for set time durations and measuring the resulting leakage current until film failure occurred.

Physical and chemical film characterizations were completed to study the surface smoothness and the chemical composition. Optical microscopy and scanning electron microscopy examined the surface topography and crystallinity of the films. Scanning Electron Microscopy - Energy Dispersive X-ray Spectroscopy (SEM-EDS) and Rutherford Backscattering Spectroscopy (RBS) were utilized to examine the chemical composition and distribution of the films and to identify impurities and contaminates in the material.

RESULTS AND DISCUSSION

Physical and Chemical Characterization

High power density capacitors are constructed by depositing thin dielectric films on flexible substrates and then rolling the substrates to manufacture the capacitors. Amorphous films are preferable to crystalline or polycrystalline films. Crystalline films are generally less flexible and grain boundaries create preferential leakage sites and shorts. Optical observations and SEM analysis of the surface reveal amorphous AlN films under the deposition conditions utilized. SEM-EDS and RBS were utilized to examine the chemical composition of the films. Figure 1 shows the EDS spectrum for a film deposited from pure nitrogen feed gas. An oxygen peak is clearly visible in the spectrum. RBS also identifies the presence of oxygen and both techniques determine ~30% oxygen in the deposited film. The composition is constant through the entire film. This oxygen may be from adsorbed gases on the deposition chamber walls. The Al:N ratio in these films is ~1:1. Increasing the DC power increases the Al concentration in the films.

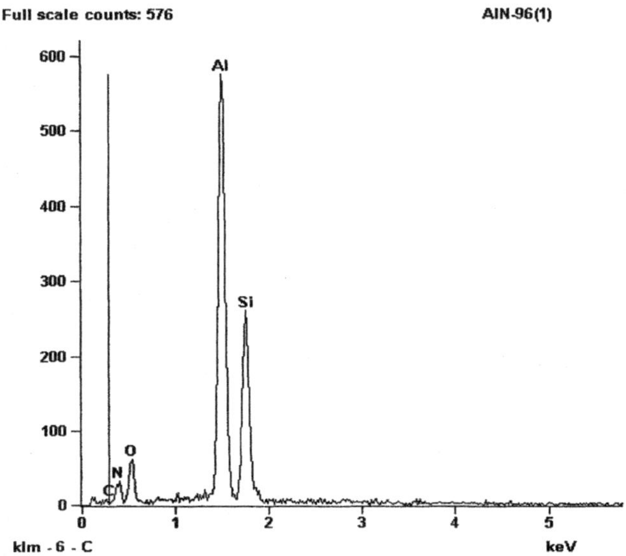

Figure 1. EDS spectrum for AlN deposited from pure nitrogen

The composition changes dramatically when oxygen is intentionally introduced into the process. Figure 2 is the EDS spectrum for a film deposited with a 10:1 ratio of N_2:O_2. The oxygen appears to completely replaces the nitrogen in the film and an AlO compound is formed. Close inspection reveals a shoulder on the oxygen peak from the presence of a small amount of nitrogen. RBS confirms the film is primarily AlO with a small amount of nitrogen, with an Al:O ratio of ~1:2, and ~7% N present. The standard heats of formation for crystalline AlN and Al_2O_3 are -318 kJ/mol and -1675.7 kJ/mol, respectively. The formation of aluminum oxide is ~5X more thermodynamically favorable than aluminum nitride. Although reactive sputtering does not operate at thermodynamic equilibrium, the heats of formation indicate any excess oxygen will displace nitrogen in the films. Formation of a stable AlON film requires tight control of the oxygen content.

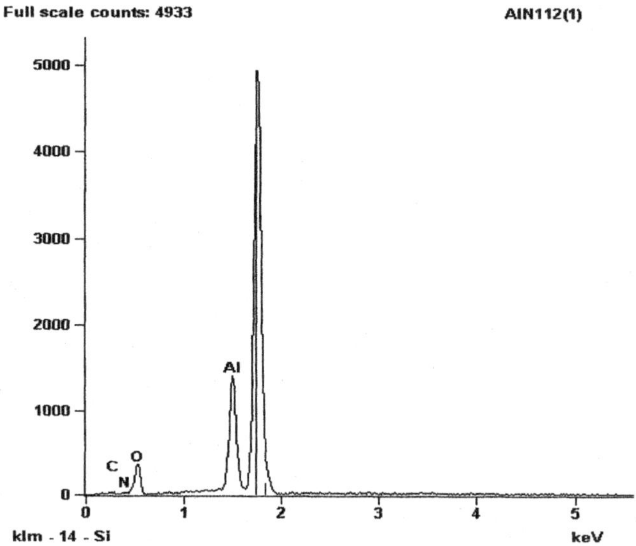

Figure 2. EDS spectrum for AlON deposited from 10:1 N_2:O_2.

Dielectric Properties

The capacitance, dissipation factor and dielectric strength were measured for AlN and AlON films. Capacitance and dissipation factor were measured as both a function of frequency from 20 Hz – 1 MHz and temperature from 20°C – 400°C. Dielectric constants for AlN films grown from pure N_2 gas calculated from the measured capacitance ranged from 7 – 10 and the dissipation factor varied from 0.01 – 0.10 at 1 kHz. These values are similar to previous studies[7]. Initial films were deposited with active substrate cooling. These films exhibited high dissipation factors and poor dielectric performance. Increasing the substrate temperature during deposition improved the dissipation factor and the active cooling was not employed for the films discussed in this work. Film properties have been compared for metallized glass, metallized silicon, aluminum foil and polymer films. No significant substrate influence was observed in the samples studied. This result differs from previous work which showed the substrate may influence crystallinity through growth kinetics[12]. Figure 3 plots the capacitance and dissipation factor vs. frequency for samples deposited from pure nitrogen and nitrogen/oxygen mixtures. The addition of oxygen significantly improves both of these film properties. The capacitance is more stable with frequency and the dissipation factor is lowered by almost on order of magnitude.

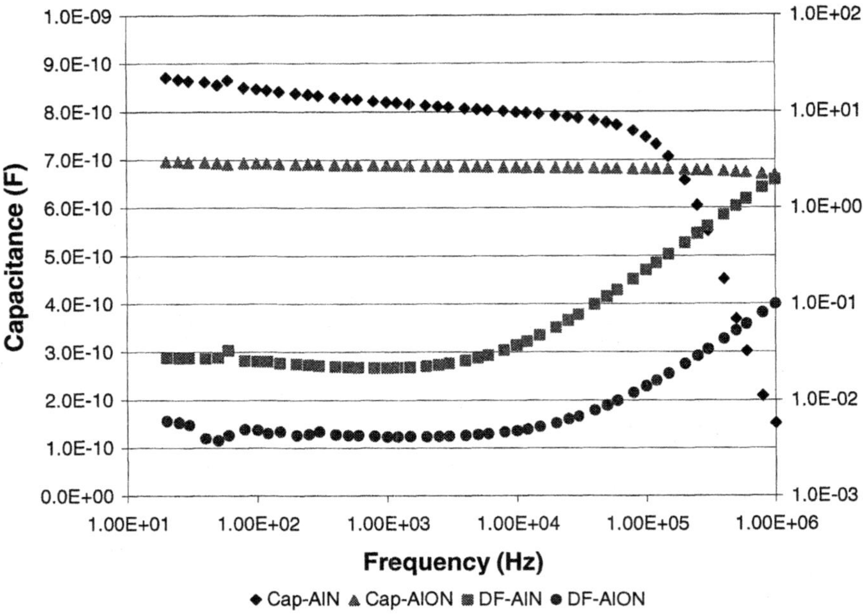

Figure 3. Capacitance and dissipation factor vs. frequency for AlN and AlON.

Temperature stability is also an important parameter for capacitor performance under extreme conditions. The capacitance is plotted vs. frequency for multiple temperatures under vacuum in Fig. 4. The capacitance remains constant with frequency and is stable with temperature to 300°C. As the temperature is increased above 300°C, the capacitance increases at low frequencies while above 10 kHz, no temperature effects are observed to 400°C. When capacitors are heated in ambient air, the capacitance at low frequency begins to increase around 200°C but still remains constant at high frequencies. The dissipation factor is more temperature dependent. Under vacuum, significant increases begin above 200°C while in air increases are observed around 150°C. The difference between film performance in vacuum and in ambient air may be related to the atmospheric boundary layer present on the surface. Interactions with particulates in the atmospheric boundary layer over the top electrode may begin to alter the electric field as the temperature increases. The boundary layer is more pronounced in atmospheric air compared to vacuum environment and the temperatures effects are seen at lower temperatures. The temperature effects on both capacitance and dissipation factor are reversible. The original values are retained after cooling the material back to room temperature and are stable after multiple temperature cycles while the same high temperature values are obtained on each cycle. This indicates the observed capacitance and dissipation shifts are not caused by a chemical reaction, but by reversible interactions between the atmosphere and the surface.

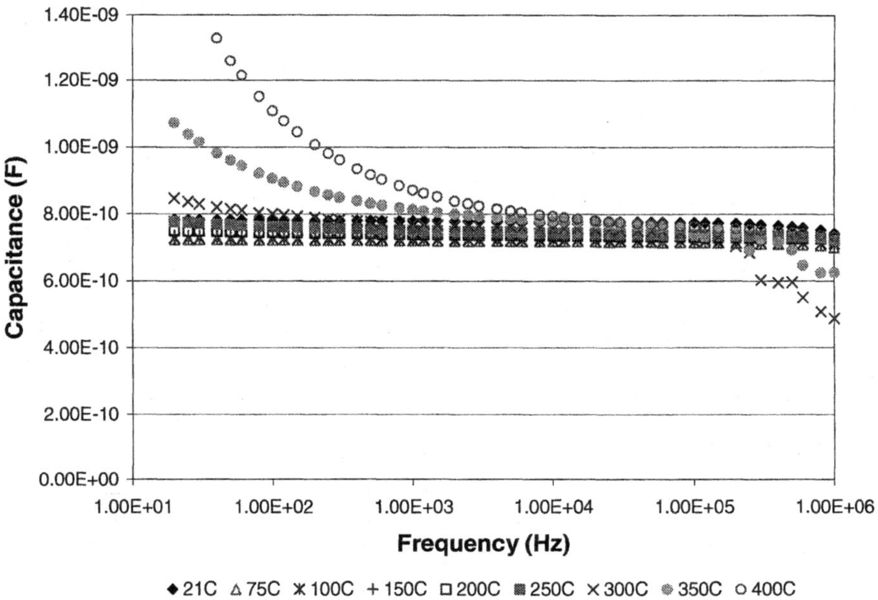

Figure 4. Effects of temperature on capacitance under vacuum.

Leakage and Breakdown

Dielectric breakdown strength is one of the most critical film properties for high power density devices. Examination of deposition parameters including DC power, deposition pressure and pulse frequency determined that DC power was the most significant input into film breakdown. Figure 5 plots leakage current vs. electric field for AlN films deposited at 2000W and 850W DC power. The film grown at lower power demonstrates higher breakdown strength. Maximum breakdown strength of ~650 V/μm has been measured. As discussed above, increased power during film growth results in a higher aluminum fraction in the material. Increased DC power has also been shown to increase micro voids in the dielectric[9]. The higher metal content or the increased void concentration creates defect sites that breakdown under lower applied voltage. Based on our optimal films, with an average dielectric constant of 8 and maximum breakdown strength of 650 V/μm, the maximum energy density of the AlON films is ~ 15 J/cc, over 50% improvement over current polymer dielectrics.

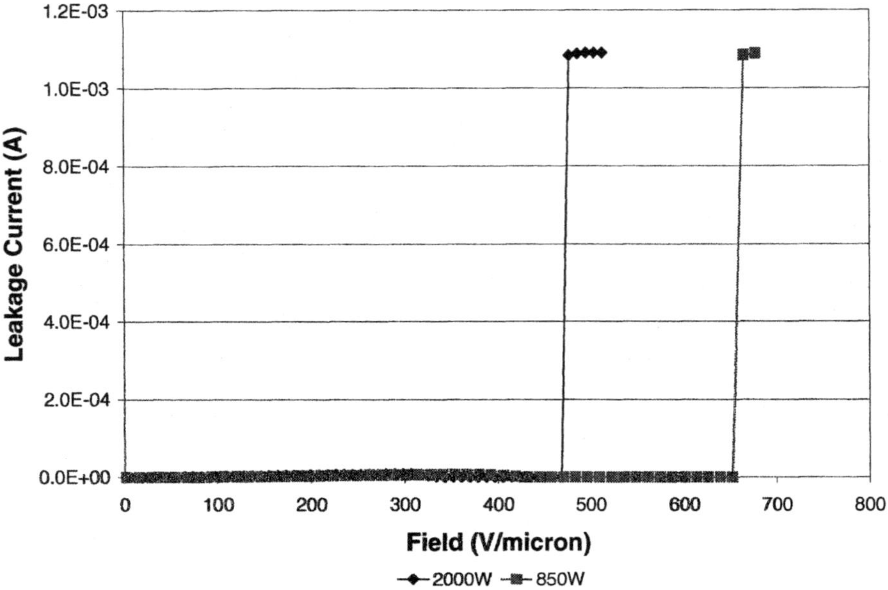

Figure 5. Dielectric breakdown strength at 2000W and 850W.

Leakage through capacitors is a thermally activated process, following the standard exponential formula:

$$I = A \exp(-E_A/kT) \tag{2}$$

where I is leakage current, A is a constant, E_A is activation energy, k is the Boltzman constant and T is absolute temperature. Knowledge of the activation energy helps predict the impact of temperature on capacitor performance. Leakage current vs. voltage measurements were taken at multiple temperatures in air, as shown in Fig. 6. The current increased with increasing temperature, consistent with a thermally activated process. Arrhenius plots of ln(I) vs. 1/T were constructed using leakage currents selected at constant voltages. The activation energy is extracted from the slope of the plot. An evaluation of E_A vs. electric field in Fig. 7 shows the activation energy is approximately constant with field with a value of 0.85±0.07eV. The capacitors will respond similarly to temperature at all fields.

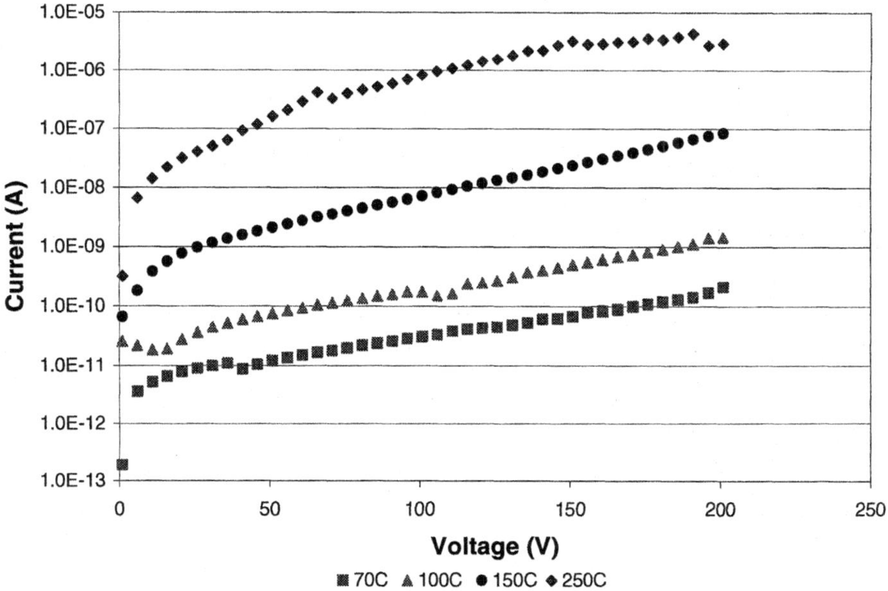

Figure 6. Leakage current as a function of temperature.

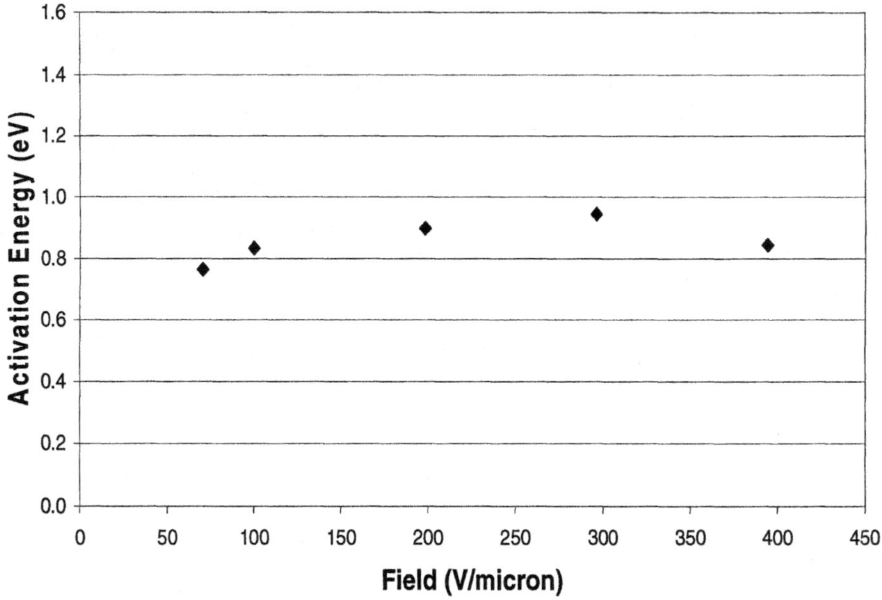

Figure 7. Activation energy vs. electric field.

CONCLUSION

Aluminum nitride and aluminum oxynitride dielectrics for high energy density capacitors applications have been deposited using pulsed DC magnetron sputtering. Films have been grown on multiple substrate materials, including aluminum foil and polymer films compatible with rolled capacitor applications. Materials properties are independent of the initial substrate. The dielectric constant and dissipation factor have been measured as a function of feed gas and measurement temperature. Dielectric constants from 7-10 have been obtained. The capacitance decreases with increasing frequency for films grown from pure nitrogen feed gas. The dissipation factor is ~0.05. The addition of oxygen to the feed gas lowers the dissipation factor to ~0.003 and increases the frequency stability of the capacitance. Temperatures above 300°C begin to affect the lower frequency range of the capacitance in vacuum. The dissipation factor is more temperature sensitive and demonstrates temperature deviations above 200°C. The temperature effects are reversible and films regain their as-deposited properties when returned to room temperature. Leakage in the AlON films is thermally activated with activation energy of 0.85eV. The activation energy is independent of the applied electric field. Dielectric breakdown strengths greater than 600 V/μm have been obtained. Increased DC power during deposition decreases the breakdown strength due to increased metal incorporation or voids in the sputtered films. Energy densities of up to 15 J/cc have been obtained on AlON films.

REFERENCES

[1]M. Rabuffi and G. Picci. "Status quo and future prospects for metallized polypropolene energy storage capacitors," *IEEE Trans. Plasma Sci.*, **30**, 1939-42 (2002).

[2]W.M. Yim, E.J. Stofko, P.J. Zanzucchi, J.I. Pankove, M. Ettenberg, and S.L. Gilbert. "Expitaxial grown AlN and its optical band gap," *J. Appl. Phys.*, **44**, 292-6 (1973).

[3]R.D. Vispute, J. Narayan, and J.D. Budai. "High quality optoelectronic grade epitaxial AlN films on α-Al_2O_3, Si, and 6H-SiC by pulsed laser deposition.," *Thin Solid Films*, **299**, 94-103 (1997).

[4]F. Martin, P. Muralt, M.-A. Dubois, and A. Pezous. "Thickness dependence of the properties of highly c-axis textured AlN thin films." *J. Vac. Sci. Technol. A*, **22**, 361-5 (2004).

[5]C.L Aardahl, J.W. Rogers Jr., H.K Yun, Y. Ono, D.J. Tweet, S.-T. Hsu. "Electrical properties of AlN thin films deposited at low temperature on Si(100)." *Thin Solid Films,* **146,** 174-80 (1999).

[6]K.K. Harris, B.P. Gila, J. Deroaches, K.N. Lee, J.D. MacKenzie, C.R. Abernathy, F. Ren, and S.J. Pearton. "Microstructure and thermal stability of aluminum nitride thin films deposited at low temperature on silicon." *J. Electrochem. Soc.,* **149**, G128-G130 (2002).

[7]V. Dimitrova, D. Manova, and E. Valcheva. "Optical and dielectric properties of dc magnetron sputtered AlN thin films correlated with deposition conditions." *Mater. Sci. Eng. B* **68**, 1-4 (1999).

[8]J. Schulte and G. Sobe. "Magnetron sputtering of aluminum using oxygen or nitrogen as reactive gas." *Thin Solid Films*, **324** (1998): 19-24.

[9]J.-W Lee and S.C.N. Cheng. "Development of SiN_x and AlN_x passivation layers." *Thin Solid Films,* **358,** 215-22 (2000).

[10]K. Jagannadham, K. Sharma, Q. Wei, R. Kalyanraman, and J. Narayan. "Structural characteristics of AlN films deposited by pulsed laser deposition and reactive magnetron sputtering: A comparative study." *J. Vac. Sci. Technol. A* **16**, 2804-15 (1998).

[11]T.T. Leung and C.W. Ong. "Nearly amorphous to epitxial growth of aluminum nitride films," *Diamond Rel. Mater.* **13**, 1603-8 (2004).

[12]F. Engelmark, J. Westlinder, G.F. Iriarte, I.V. Katardjiev, and J. Olsson. "Electrical characterization of AlN MIS and MIM structures," *IEEE Trans. Electr. Dev.*, **50**, 1214-9 (2003)

HIGH TEMPERATURE PIEZOELECTRIC $La_2Ti_2O_7$

Ali Sayir, Serene C. Farmer and Fred Dynys

NASA Glenn Research Center
Cleveland, Ohio 44135

ABSTRACT

$La_2Ti_2O_7$ belongs to a family of ferroelectric materials having a layered perovskite structure. Its unusually high Curie temperature and unique layered structure makes investigation of $La_2Ti_2O_7$ of interest for high temperature piezoelectric devices. High temperature properties of $La_2Ti_2O_7$ were measured on single crystals and sintered samples with theoretical densities of 95 %. Severe cleavage fracture made handling of single crystals difficult. Particular attention was devoted to identifying and controlling the processing parameters that influence oxygen stoichiometry and piezoelectric properties. Sintered $La_2Ti_2O_7$ is an insulator exhibiting resistivity values on the order of 10^{11} ohm-cm at room temperature. The resistivity decreased by 5 orders of magnitude at 650 °C, a secondary grain boundary phase was responsible for the decrease in resistivity at elevated temperatures. Means to eliminate the second phase was investigated. The strain as a function of applied electric field strength was found to be independent of the field direction, indicative of electrostrictive behaviour. Values up to 0.4 % strain were measured. The total strain was dependent upon the electric field strength. The effect of Nd –doping on polarization was examined as a function of electric field. Nd-doping on the La site increased the polarization significantly indicating the importance of La-site doping.

INTRODUCTION

The aeronautics and aerospace applications require high performance electronics, power conditioning systems, sensors and actuators that can operate reliably in extreme environments. Placement of the actuators on the hot surfaces such as casing stator vanes will require piezoceramic devices compatible with surface temperatures ranging from 300 – 600 °C [1, 2]. The temperature requirement is significantly greater than the maximum use temperature for commercial piezoelectric materials which is typically 80 – 140 °C. There is a need to develop higher temperature piezoceramic materials and demonstrate sensing and actuation capabilities at elevated temperatures. Establishing high temperature piezoelectric properties of new sensing/actuating materials is the primary focus of the NASA GRC effort. High temperature piezoelectric actuators and sensors have potential to reduce engine emissions and improve efficiency.

The application temperatures for most piezoelectric materials are limited by their Curie temperature, T_c. Performance degradation through domain re-orientation (depoling) occurs at temperatures greater than $T_c/2$. The family of high T_c ferroelectrics with perovskite-like layer structures (PLS), sometimes referred to as the strontium pyriobate family, were found to exhibit strong piezoelectric effect at ambient temperatures. $La_2Ti_2O_7$, $Nd_2Ti_2O_7$ [6, 7] and alkali-metal niobates, such as $Sr_2Nb_2O_7$ [8, 9] are isomorphous and have layered structures belonging to homologous series $A_{n+1}B_{n+1}O_{3n+5}$, $0 \leq n \leq \infty$. $A_2B_2O_7$ ($A_4B_4O_{14}$) structures are high transition-temperature ferroelectrics, $T_c > 1500$ °C. Kimura [6] determined the piezoelectric properties of $La_2Ti_2O_7$ single crystals along the polar axis (<010> direction of the monoclinic crystal (space group $P2_1$) and reported respectable properties. The electromechanical coupling factor k_{22}= 0.29, piezoelectric constant d_{22} = 16 x10^{-12} C/N and elastic compliance s_{22}^E = 5.5x10^{-12} m^2/N have been measured at room temperature. In a follow up study, the piezoelectric properties of single crystal $Nd_2Ti_2O_7$ was determined to be k_{22} = 0.15, d_{22} = 6.5 x10^{-12} C/N and s_{22}^E = 4.8x10^{-12} m^2/N [7]. Both $La_2Ti_2O_7$ and $Nd_2Ti_2O_7$ exhibit Curie temperatures greater than 1500 °C. Both materials are candidates for high temperature applications. Nanamatsu et al. [10] studied the *D – E hysteresis* loop of $La_2Ti_2O_7$ single crystals at room temperature and determined the spontaneous polarization P_s and the coercive field E_c. The coercive field was almost saturated ($E_c = 45\ kV/cm$) but spontaneous polarization did not saturate, P_s = 5 µC/cm^2. There are also significant disadvantages of these perovskite-like layer structures. The low crystal symmetry of $La_2Ti_2O_7$ makes aligning domains difficult and cleavage fracture occurs along the {100} planes. Materials will be investigated with these limitations in mind.

In this investigation, we experimentally determined the phase formation characteristic of $La_2Ti_2O_7$ from the reaction of mixed oxide phases, La_2O_3 and TiO_2. The aim was to produce dense ceramics with high electrical resistance and to measure the temperature limitations of $La_2Ti_2O_7$. Phase analysis using x-ray diffraction is reported with the emphasis on phase determination at compositions near the line compound $La_2Ti_2O_7$ in the La_2O_3 and TiO_2 system [11, 12, 13]. The displacement as a function of electric field was measured directly to assess the strain capability. Nd –doping on the La- site of the $La_2Ti_2O_7$ composition was investigated to modify the polarization characteristic of the $A_2B_2O_7$ ($A_4B_4O_{14}$) structure by recording polarization versus electric field ($P – E$) hysteresis loops.

EXPERIMENTAL

The raw materials were high purity La_2O_3 (99.999 % Aithica Chemical Corp.) and TiO_2 (99.99 %, Alfa Aesar – Johnson Matthey). Source materials were milled for several days in teflon containers in ethy lalcohol without milling agents. The alcohol was evaporated, and powder was pressed into 25 mm in diameter pellets at a uniaxial pressure of 130 MPa. For single crystal growth, source rods (10 mm in diameter) were fabricated by cold isostatic pressing using a hydrostatic pressure of 400 MPa. Source rods were sintered at 1450 °C in air. Platinum electrodes were fabricated from platinum paste (Heraeus CL11-5349) by firing at 1050 °C for one hour in air.

Single crystal growth was achieved using the laser heated float zone method [14]. The CO_2 laser beam was split into two beams, 180 ° apart from each other. Each beam was then focused with zinc-selenide lenses to heat and form a molten zone at the top of a polycrystalline source rod. Both the source rod and the solidified rod moved vertically through the laser beam using a pulling rate of 2 mm/h. Most of the radiation stemming from CO_2-laser (wavelength 10.6 µm) were absorbed by the oxide. The controllable growth parameters include the molten zone height, source rod to crystal diameter ratio, pull-rate, wetting angle and laser power distribution. In-depth description of experimental conditions has been described in previous work [14].

The crystalline phases of the calcined and sintered specimens were characterized by x-ray diffraction (XRD) and the microstructures examined by scanning electron microscope (SEM). The x-ray diffractometer was equipped with a Cu K_{α} source with a wavelength of 0.1540 nm. The x-ray scans were conducted at 3°/min with a sampling interval of 0.02°.

The electrical properties as a function of the frequency and temperature were determined by electrical impedance spectroscopy (EIS). Solartron 1260 impedance analyzer/Solartron 1287 electrochemical interface or Solartron 1260/1296 dielectric interface were used to acquire impedance spectrum from 0.1 to 1 MHz. Samples were heated in an air environment and allowed to equilibrate for 1 hr. before testing. The complex impedance plots were analyzed by fitting with equivalent circuits to obtain the resistance and capacitance as a function of temperature. The total conductivity was calculated from the impedance data and sample geometry. The equivalent circuit $(RQ)_{Grain}(RQ)_{GrainBoundary}(R)_{electrode}$, where R = resistance and Q = constant phase element or CPE, was used to model the impedance spectrum. Circuit parameters for the equivalent circuits were obtained by using the least squares fitting routine in Zview (Scribner Aossiciates).

Samples were poled in air in a custom made mini-furnace with SiC heating elements. The dielectric behavior was measured using a multifrequency LCR meter (Agilent 4284A). Attempts were made to identify resonance and antiresonance frequencies using impedance analyzer (HP4194A). Polarization characteristics were determined at high electric fields using a modified Sawyer-Tower circuit (Radiant Technology). Strain measurements were made using the MTI-200 Fotonic sensor which uses the principle of optical lever to resolve changes of less than 0.1 μm in sample displacement. The Fotonic system uses step-index fiber optics, where transmission fibers illuminate a surface and receiving fibers transfer the reflected light to a photocell. Changes in displacement cause the photocell to detect an amplitude oscillation of the reflected light, which can be converted into displacement after calibration.

RESULTS AND DISCUSSION

The crystalline solid solutions explored were fabricated using conventional mixed oxide ceramic powder processing. Through x-ray diffraction analysis, it was determined that as little as 1 wt % excess Ti promoted the formation of $La_2Ti_3O_9$ as a second phase. The compound $La_2Ti_3O_9$ is a defect perovskite containing up to 1/3 cation vacancies on the La site, commonly expressed with the formula of $La_{2/3}TiO_3$ [15]. $La_{2/3}TiO_3$ exhibits increased electronic and ionic conductivity [15 - 18]. The formation of $La_{2/3}TiO_3$ as a second phase in $La_2O_3 - TiO_2$ system reduces the electrical resistivity, rendering the $La_2Ti_2O_7$ material useless for piezoelectric applications. Precision controlled processing procedures are required to achieved phase pure $La_2Ti_2O_7$. Starting materials must be weighed with precision. La_2O_3 readily reacts with water and CO_2 during storage. To remove the physically and chemically reacted water from the La_2O_3 powders, a two-step calcination process was adopted as suggested by Bernal et al. [19]. The two step process required calcination of the La_2O_3 powder for 10 h at 600 and 800 °C. The calcination removed the physically adsorbed and chemically bound water. The powder was weighed immediately (within 5 minutes) following the calcination to ensure an accurate La_2O_3 to TiO_2 ratio. This approach ensured accurate stoichiometry of starting compositions and formation of phase pure $La_2Ti_2O_7$.

Reaction conditions used to form $La_2Ti_2O_7$ was also important for eliminating the formation of undesirable phases. Reaction temperature and environment were changed to find the process conditions which form $La_2Ti_2O_7$. At 1400 °C in flowing O_2, a polyphase mixture containing $La_4Ti_9O_{24}$, $La_2Ti_2O_7$ and Ti-rich phase was detected. Further reaction time at 1400 °C produced a binary mixture of $La_2Ti_2O_7$, Ti-rich phase or La_2TiO_5. Scanning electron microscopy (SEM) analysis revealed a Ti-rich grain boundary phase between the $La_2Ti_2O_7$ and La_2TiO_5 phases. Multiple calcinations at 1400 °C could not yield single phase $La_2Ti_2O_7$, La_2TiO_5 remained present as inclusions. The presence of La_2TiO_5 phase and its persistence at very high temperatures was not reported by Fuierer et al. [20] and Zhong et al. [21]. Even though both of these investigations involved sintering of $La_2Ti_2O_7$ using conventional mixed-oxide processing. In the present investigation, single-phase $La_2Ti_2O_7$ samples were successfully fabricated by sintering compacts at temperatures in excess of 1500 °C in flowing oxygen, Fig. 1. The flowing O_2 atmosphere was used to eliminate the formation of the n = 4 compound (e.g., $La_5Ti_5O_{17}$) as suggested by Williams et al. [17]. The experimental results of our investigation for the phase formation characteristics in the $La_2O_3 - TiO_2$ system agree with the recent phase diagram results of Skapin et al. [13]. However, we observed the formation of $La_2Ti_2O_7$ from the $La_4Ti_9O_{24}$ plus Ti-rich melt at a lower temperature (1400 °C) than Skapin et al. This is not surprising since La_2O_3-TiO_2 system is not a true binary system. The phase diagram is to be regarded as a section at partial oxygen pressure of air of 1 atm.

The electrical properties of the phase pure $La_2Ti_2O_7$ samples were measured by EIS. A series of representative spectra over the complete temperature range (25 – 600 °C) are shown on a Nyquist plots in Fig. 2. Two distinct impedance arcs were observed. The large arcs at the higher frequencies have been attributed to the response of grain interiors, while the smaller arcs have been attributed to the response of grain boundaries. Only one arc was observed at temperatures less than 150 °C. Grain boundary conduction by second phases is expected to dominate at higher temperatures. As the temperature increased, the resistance decreased and the time constants *(RC)* of the relaxations associated with polarizations have been reduced shifting the arcs to higher frequencies. The data shows that grain boundary resistance is smaller than the grain resistance. An increase in grain boundary conduction can originate from a departure of stoichiometry that can be accommodated by the $A_{n+1}B_{n+1}O_{3n+5}$ structure. Williams et al [17] used thermogravimetric and high-resolution transmission electron microcopy (HRTEM) study to illustrate that the structure tolerates a wide range of compositions, specifically at grain boundaries. In addition, possible formation of n = 4 compound (e.g., $La_5Ti_5O_{17}$) as a grain boundary phase cannot be ruled out. The exact mechanism for the grain boundary conduction is currently not known and will require further study.

This work showed that conventional calcining and milling of mixed-oxide sintering of $La_2Ti_2O_7$ could generate resistivities up to 10^{11} ohm cm for samples with theoretical densities of 95 %. An additional increase in resistivity could be achieved by sintering to full density or by minimization of grain boundary phase conduction. The amount of grain boundary phases may define the upper use temperature of $La_2Ti_2O_7$ as a piezoelectric ceramic. It is important to note that the conventional processing approach (calcining and milling of mixed-oxides) may be a preferred means to produce high resistivity $La_2Ti_2O_7$ ceramics. Alternative processing routes which

produce fine grains [20, 22] have an associated large amount of grain boundaries and may be less desired for piezoelectric applications unless grain boundary phases can be eliminated or reduced.

Successful poling requires that the ferroelectric material be a good insulator with high dielectric breakdown strength. Observing the polarization (*P*) versus applied electric field strength (*E*), $La_2Ti_2O_7$ exhibited only a modest level of polarization with applied fields. The lack of saturation polarization (P_s) is an indication for the absence of domain growth. Our efforts to pole the polycrystalline $La_2Ti_2O_7$ at electric field levels up to 90 kV /cm and at temperatures up to 600 °C were not successful. The low crystal symmetry ($P2_1$) of $La_2Ti_2O_7$ made it difficult to pole the randomly oriented polycrystalline ceramic. The dipole moment in $A_{n+1}B_{n+1}O_{3n+5}$ structure is generated by the displacement of the interlayer through a cation displacement in the (100) cleavage plane that is parallel to the b-axis (polar axis) [8, 9]. Although we do have very little understanding of the poling mechanism [23], we doped the $La_2Ti_2O_7$ with Nd cations in an effort to improve the dipole moment per unit volume. Figure 4 includes a polarization versus applied electric field of the undoped and Nd-doped $La_2Ti_2O_7$ for comparison. The microstructure of $(La_{0.8}Nd_{0.2})_2Ti_2O_7$ samples contained higher porosity than $La_2Ti_2O_7$, but still the level of polarization and coercive field increased with Nd addition. This experiment emphasizes the significance of A-site cations in the structure. Kim et al. [25] have shown that thin films of $Nd_2Ti_2O_7$ have significant remnant polarization (P_c = 4. 4 μC / cm^2) and coercive field (E_c = 131 kV/cm^2). These values are noteworthy in view of the fact that these measurements were made on polycrystalline films derived by sol-gel method. The remnant polarization values of these films are an order of magnitude higher than the P_c values measured in the present investigation (Fig. 4), leaving room for further materials development. In addition, the ferroelectric hysteresis loop of the sol-gel derived films showed a significant degree of saturation polarization (P_s) characteristic as a result of maximum domain growth. Additional off-centre displacement of Ti atoms in $[TiO_6]$ octahedral and dipoles can possibly be transmitted to neighboring actahedra through corner sharing octahedral network. Titov et al. [24] proposed different ordering PLS slabs through engineering of *B-site* of the $A_{n+1}B_{n+1}O_{3n+5}$ structure and hypothesize possible increase of spontaneous polarization, P_s. Further experimental study is required to understand the polarization mechanism through both A-, and B-site dopants of the $A_{n+1}B_{n+1}O_{3n+5}$ structures.

Kimura et al. [6, 7] and Nanamatsu et al. [10] have shown spontaneous polarization of 5 and 10 μC/cm^2 for single crystals of $La_2Ti_2O_7$ and $Nd_2Ti_2O_7$, respectively. They did not report any attempt to pole the single crystals. We were successful in growing several single crystals of $La_2Ti_2O_7$ composition using a laser heated melt modulation technique. [14] Fig. 5 shows a representative sample. However, strong cleavage characteristic of the single crystals made handling impractical.

The polycrystalline $La_2Ti_2O_7$ material was a good insulator with high dielectric breakdown strength and consequently we studied the actuation at high field levels. The strain as a function of applied electric field strength was independent of the field direction *(+E or –E)*. The total strain was found to be dependent on the electric field. The material exhibits an *electrostrictive effect.* Figure 6 shows an electric-field induced strain of $La_2Ti_2O_7$ material with large grains (>30 μm). The amount of displacement is significant. Extended sintering times at high temperatures produced large $La_2Ti_2O_7$ grains, possibly producing a near single domain state. This is in agreement with previous report that $La_2Ti_2O_7$ single crystals grown by the float zone technique along the c-axis were nearly single domain [23]. The processing techniques that can create single-domain $La_2Ti_2O_7$ could eliminate tedious and difficult poling process. This provides an opportunity to develop these materials for high temperature sensors or actuators and subject of our current research efforts.

SUMMARY

Phase pure $La_2Ti_2O_7$ can be produced by conventional powder processing following procedures which remove the physically and chemically reacted water before weighing reactants and then sintering compacts at temperatures in excess of 1500 °C in flowing oxygen. Electrical resistance and impedance values obtained as a function of temperature on polycrystalline $La_2Ti_2O_7$ indicated strong grain boundary contribution. Attempts to pole polycrystalline $La_2Ti_2O_7$ up to 600 °C using 90 kV/cm were not successful. The substitution of La atoms with Nd atoms in the A site of $A_{n+1}B_{n+1}O_{3n+5}$ structure increased polarization but also high losses. $La_2Ti_2O_7$ showed large strain under large electric fields independent of the field direction.

REFERENCES

[1] W. K. Lord, D. G. McMartin and T. G. Tillman, " Flow Control Opportunities in Gas Turbine Engines," AIAA (2000) - 2234.

[2] D. C. McCormic, "Boundary Layer Separation Control with Directed Synthetic Jets," AIAA (2000) – 0519.

[3] J. M. Cohen, J. H. Stufflebeam, and W. Proscia, "The Effect of Fuel/Air Mixing on Actuation Authority in an Active Combustion Instability Control System," J. Engineering For gas Turbine and Power, 123 (2001) 537 – 544.

[4] R. H. Thomas, M. M. Choudhari and R. D. Joslin, "Flow and Noise Control: Review and Assessment of Future Directions," NASA / TM-2002 -211631.

[5] A. Scotti, C. Meneveau and S. Saddoughi, "Fractal Dimension of Velocity Signals in High-Reynolds-Number Hydrodynamic Turbulence," Phy. Rev. E, 51[6] (1995) 5594 -5601.

[6] M. Kimura, S. Nanamatsu, K. Doi, S. Matsushita, and M. Takahashi, "Electrooptic and Piezoelectric Properties of $La_2Ti_2O_7$ Single Crystal," Jpn. J. Appl. Phys., 11 (1972) 904.

[7] M. Kimura, S. Nanamatsu, T. Kawamura and S. Matsushita, "Ferroelectric, Electrooptic and Piezoelectric Properties of $Nd_2Ti_2O_7$ Single Crystal," Jpn. J. Appl. Phys., 13 (1974) 1473 -1474.

[8] J. K. Brandon and H. D. Megaw, "On the Crystal Structure and Properties of $Ca_2Nb_2O_7$ 'Calcium Pyroniobate'," Phil. Mag., 21 [8] (1970) 189 – 194.

[9] N. Ishizawa, F. Marumo, S. Iwai, M. Kimura and T. Kawamura, "Compounds with Perovskite-Type Slabs, V. A High-Temperature Modification of $La_2Ti_2O_7$," Acta Crystallogr., B38 (1982) 368 - 372.

[10] S. Nanamatsu, M. Kimura, K. Doi, S. Matsushita and N. Yamada, "A New Ferroelectric: $La_2Ti_2O_7$,"Ferroelectrics, [8] (1974) 511 – 513.

[11] R. Roth, ""Pyrochlore-Type Compounds Containing Double Oxides of Trivalent and Tetravalent Ions," J. Res. of the Natl. Bureau of Standards, 56 [2643] (1956) 17 – 25.

[12] J. B. MacChesney and H. A. Sauer, "The System La_2O_3 – TiO_2; Phase Equilibria and Electrical Properties," J. Am. Ceram. Soc., 45 [9] (1962) 416 422.

[13] S. D. Skapin, D. Kolar and D. Suvorov, "Phase Stability and Equilibria in the $La_2O_3 - TiO_2$ System," J. Europ. Ceram. Soc., 29 (2000) 1179 – 1185.
[14] A. Sayir, S. C. Farmer, Pat Dickerson, and Arnon Chait, "Laser Heated Oxide Fiber Growth Using Melt Modulation Technique," in Mtg. Proc. 9-Intelligent Processing of High Performance Materials - RTO-MP-9 AC/323 (AVT) TP/5 (1998) 17-1-17.17.
[15] M. Kestigan and R. Ward, "The Lanthanum – Titanium – Oxygen System," J. Am. Ceram. Soc., 77 (1955) 6199 – 6170.
[16] M. Abe and K. Uchino, "X-Ray Study of the Deficient Perovskite $La_{2/3}TiO_3$," Mat. Res. Bull., 9 (1974) 147 -156.
[17] T. Williams, H. Schmalle, A. Reller, F. Lichtenberg, D. Widmer and G. Bednorz, "On the Crystal Structures of $La_2Ti_2O_7$ and $La_5Ti_5O_{17}$: High – Resolution Electron Microscopy," J. Solid State Chem., 93 (1991) 534 – 548.
[18] T. Williams, F. Lichtenberg, D. Widmer, G. Bednorz and A. Reller, "Layered Perovskitic Structures in Pure and Doped $LaTiO_{3.5-x}$ and $SrNbO_{3.5-x}$," J. Solid State Chem., 103 (1993) 375 – 386.
[19] S. Bernal, F. J. Botana, R. García and J. M. Rodríguez-Izquierdo, "Thermal Evolution of a Sample of La_2O_3 Exposed to the Atmosphere," Thermochimica Acta, 66 [1-3] (1993) 139 – 145.
[20] P. A Fuierer and R. E. Newnham, "$La_2Ti_2O_7$ Ceramics," J. Am. Ceram. Soc., 74 [11] (1991) 2876 – 2881.
[21] Y. J. Zhong, F. Azough and R. Freer, "The Effect of $La_2Ti_3O_9$ Second Phase on the Microstructure and Dielectric Properties of $La_2Ti_2O_7$ Ceramics," J. Europ. Ceram. Soc, 15 (1995) 255 – 263.
[22] M. M. Milanova, M. Kakihana, M. Arima, M. Yashima and M. Yoshimura, "A Simple Solution Route to the Synthesis of Pure $La_2Ti_2O_7$ and $Nd_2Ti_2O_7$ at 700 -800 °C by Polymerized Complex Method," J. Alloys and Compounds., 242 (1996) 6 – 10.
[23] J. K. Yamamoto and A. S. Bhalla, "Piezoelectric Properties of Leyered-Perovskite $A_2Ti_2O_7$ (A = La and Nd) Single Crystal Fibers," J. Appl. Phys., 70 [8] (1991) 4469 – 4471.
[24] Yu A. Titov, A. M. Sych, V. Ya. Markiv, N. M. Belyavina, A. A. Kapshuk and V. P. Yaschuk, "Synthesis, Structure and Ferroelectric Properties of Heteroblock Slab Phases in $La_2Ti_2O_7 - LaTaO_4$ System," J. Alloys and Compounds, 316 (2001) 309 – 315.
[25] W. S. Kim, S-M. Ha, J-K. Yang and H-H. Park, "Ferroelectric-Gate Field Effect Transistors using $Nd_2Ti_2O_7$/ Y_2O_3/Si Structures," Thin Solid Films, 398-399 (2001) 663 – 667.

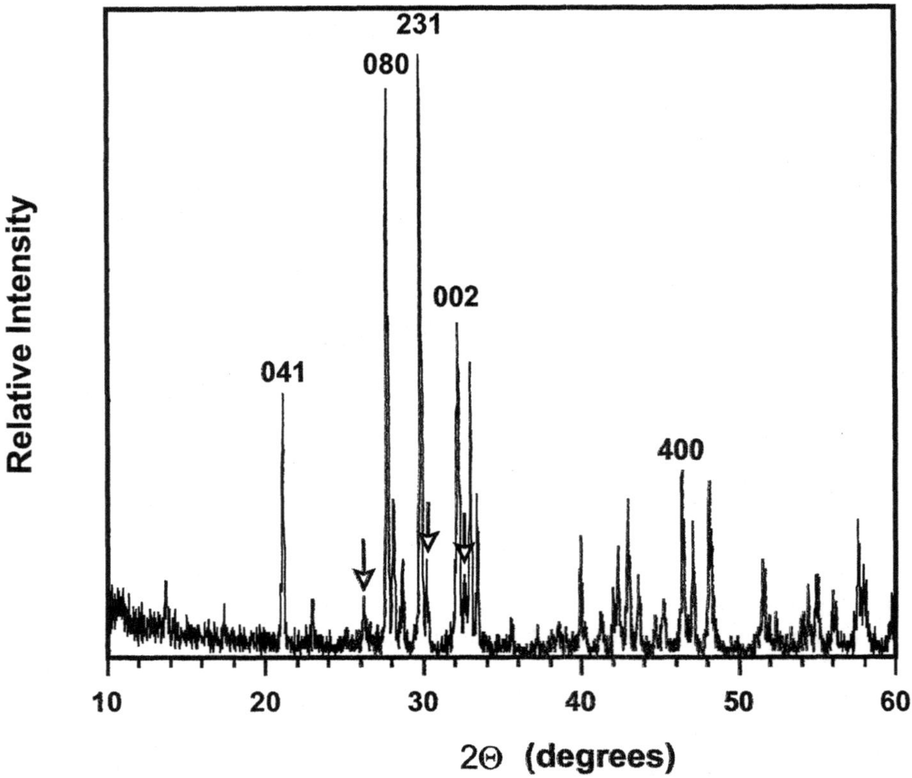

Fig. 1 X-ray diffraction powder patterns of $La_2Ti_2O_7$. Two unknown peaks are marked with arrow. Annealing times exceeding 10 hours at 1500 °C yields phase pure $La_2Ti_2O_7$.

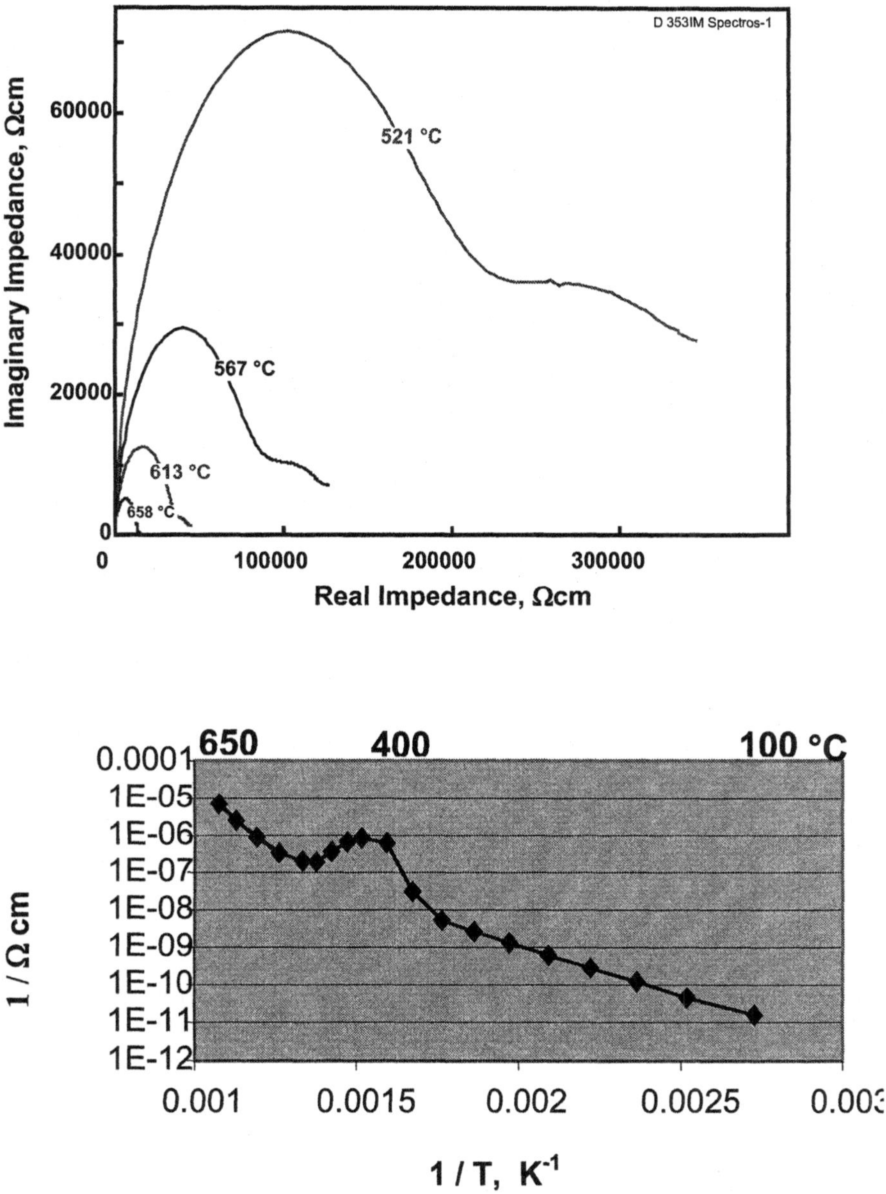

Fig. 2 Nyquist plots of $La_2Ti_2O_7$ as function of temperature (top). Total conductivity of $La_2Ti_2O_7$ as function of temperature (bottom).

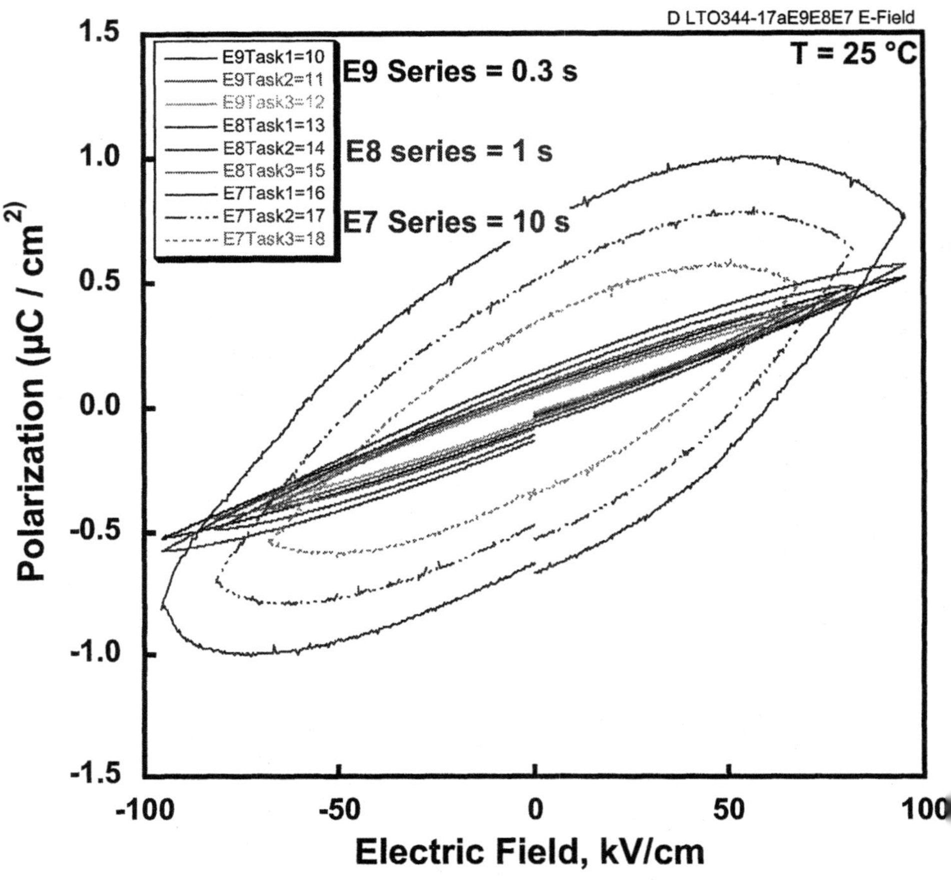

Fig. 3 Polarization as a function of electric field. For $La_2Ti_2O_7$ red and blue curves with small amount of remnant polarization. For Nd-doped $La_2Ti_2O_7$ increased remnant polarization (green curves).

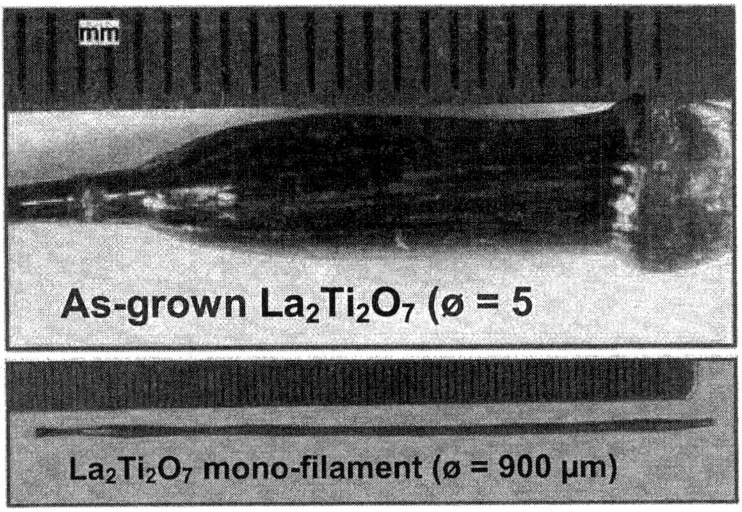

Fig. 4 single crystals of $La_2Ti_2O_7$.

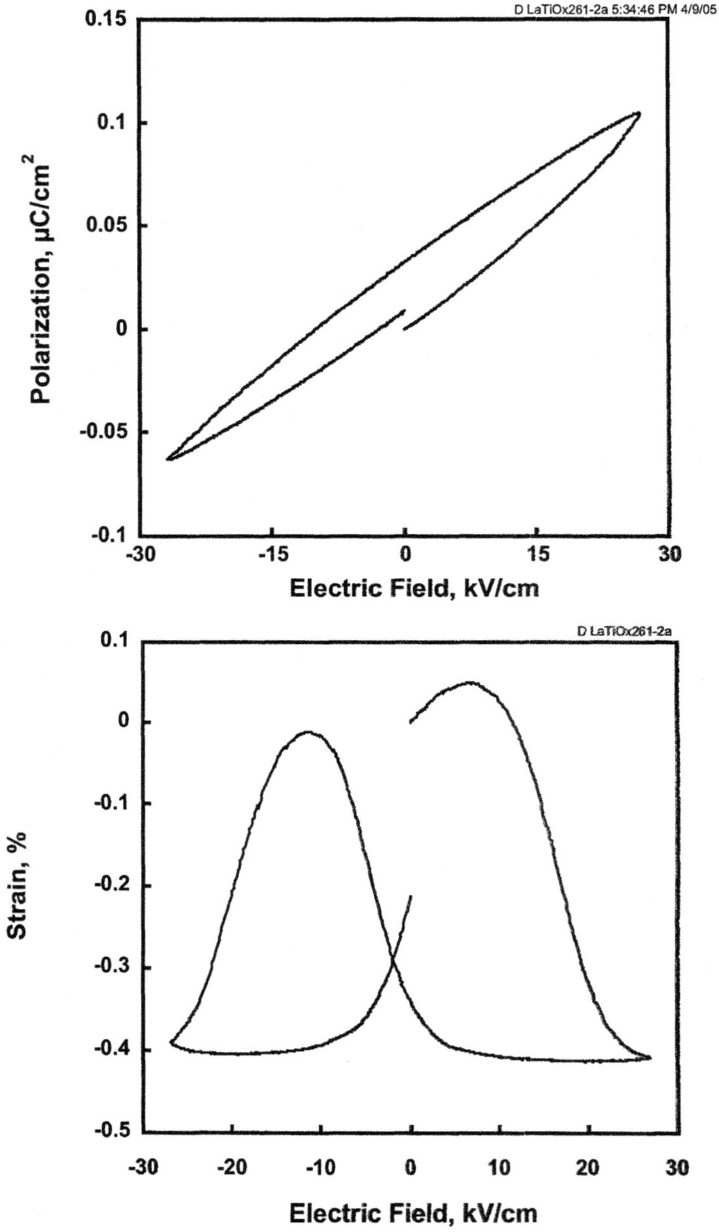

Fig. 5 Polarization (left) and strain capabilty (top) as a function of Electric field at room temperature (bottom).

THERMOPHYSICAL PROPERTIES OF PEROVSKITE TYPE ALKALINE EARTH HAFNATES

Takuji Maekawa, Ken Kurosaki, Hiroaki Muta, Masayoshi Uno, Shinsuke Yamanaka
Department of Nuclear Engineering, Graduate School of Engineering, Osaka University, 2-1, Yamadaoka, Suita, Osaka, 565-0871, Japan

Tetsushi Matsuda, Shin-ichi Kobayashi
Nuclear Fuel Industries, Ltd., 950, Ohaza-Noda, Kumatori-cho, Sennan-gun, Osaka, 590-0481, Japan

ABSTRACT

Polycrystalline samples of the perovskite type alkaline earth hafnates, $BaHfO_3$ and $SrHfO_3$ were prepared by mixing the appropriate amounts of HfO_2, $BaCO_3$ and $SrCO_3$ followed by reacting at 1273 K in air and sintering. The crystal structure and lattice parameter were evaluated with an X-ray diffraction analysis. The thermal expansion coefficient, melting temperature, heat capacity, thermal diffusivity, thermal conductivity, elastic modulus, Debye temperature, and micro-Vickers hardness were measured. The crystal structure of $BaHfO_3$ is a cubic perovskite with the lattice parameter of 0.4171 nm at room temperature. The average linear thermal expansion coefficients are 7.97×10^{-6} K^{-1} for $BaHfO_3$ and 11.3×10^{-6} K^{-1} for $SrHfO_3$ in the temperature range between 300 K and about 1200 K. The melting temperatures of $BaHfO_3$ and $SrHfO_3$ are 2893 and 3200 K, respectively. The Young's modulus of $BaHfO_3$ is 194 GPa (91 %T.D.), which is slightly lower than that of $SrHfO_3$ (220 GPa, 91 %T.D.). The thermal conductivities of $BaHfO_3$ and $SrHfO_3$ at room temperature are 14.1 and 5.20 $Wm^{-1}K^{-1}$, respectively.

INTRODUCTION

The perovskite type oxides have potential to be attractive functional materials because they have various unique properties in spite of their relatively simple crystal structure. For example, $SrTiO_3$ is a typical perovskite dielectric material with a wide range of technological applications. Its special properties, such as ferroelectricity [1], semiconductivity [2], superconductivity [3], catalytic activity [4], and thermoelectricity [5], have been reported extensively.

On the other hand, thermal barrier coatings (TBCs) of partly Y_2O_3 stabilized ZrO_2 (PYSZ) films are widely used to protect hot section parts of aircraft and land-based turbines by reducing the temperature of metal substrates. Their continued development is essential for improving the efficiency and performance of gas turbines by allowing the inlet gas temperature to be increased further. However this standard material has a limited temperature capability due to accelerated sintering and phase transformations at high temperatures. As a result, a world-wide effort has been undertaken to identify new candidates for a TBC application [6,7]. In general, TBC materials have to fulfill most of the following requirements, such as a stable phase, low thermal conductivity (<2 $Wm^{-1}K^{-1}$), high thermal expansion coefficient ($>9 \times 10^{-6}$ K^{-1}), chemical resistance, low sintering rate and high fracture toughness.

$BaHfO_3$ and $SrHfO_3$ have been well known for a long time, and it has a high melting temperature [8,9]. Recently, Kennedy et al. [10] investigated the high temperature phase transitions using powder neutron diffraction for $SrHfO_3$.

We have systematically studied the thermophysical, mechanical, and thermoelectric properties of barium and strontium series perovskite type oxides, $BaUO_3$, $BaZrO_3$, $BaCeO_3$, $BaMoO_3$, $SrCeO_3$, $SrZrO_3$, $SrRuO_3$, and $SrHfO_3$ [11-20]. In the present study, the thermophysical properties, viz. the thermal expansion coefficient, melting temperature, elastic moduli, Debye temperature, Vickers hardness, heat capacity, and thermal conductivity of $BaHfO_3$ and $SrHfO_3$ were measured.

EXPERIMNTAL PROCEDURE

The polycrystalline samples of $BaHfO_3$ and $SrHfO_3$ were prepared by mixing the appropriate amounts of HfO_2 (Furuuchi Chemical, 99.9 %), $BaCO_3$ (Wako Chemical, 99.9 %), and $SrCO_3$ (Aldrich Chemical, 99.9+ %), followed by reacting at 1273 K in air. After that, the sample of $SrHfO_3$ was sintered at 1873 K for 10 h in air. On the other hand, the $BaHfO_3$ powder was placed into a 20 mm diameter graphite die and given a spark plasma sintering (SPS, SUMITOMO COAL MINING Dr Sinter SPS-1020 apparatus) at 1773 K under argon atmosphere. The temperature was measured by means of an optical pyrometer focused on the graphite die, which centered on the sintering sample. During heating and soaking, a pressure of 50 MPa was applied to the sample. The heating rate was 300 K min^{-1} from room temperature to 873 K and 100 K min^{-1} from 873 K to 1773 K. The powder X-ray diffraction (RIGAKU RINT2000) analysis was performed at room temperature using Cu-Kα radiation. In order to determine the chemical composition, the SEM-EDX analysis was performed using HITACHI S-2600H SEM instrument equipped with the energy-dispersive HORIBA EX-200 system. The oxygen analysis was performed using HORIBA EMGA-550 to identify the oxygen concentration of the samples. For measurements of the thermophysical properties, appropriate shapes of the samples were cut from the sintered pellets. The bulk density of the samples was determined by the geometric measurement.

The high temperature powder X-ray diffraction analysis was performed from room temperature to about 1500 K to study the thermal expansion behavior. The melting temperature was measured by the thermal arrest method under a reducing atmosphere. The longitudinal and shear sound velocities were measured by the ultrasonic pulse-echo method (NIHON MATECH Echometer 1062) at room temperature in air. The sintered sample was bonded to a 5 MHz longitudinal or shear sound wave echogenic transducer. From the sound velocities, the elastic moduli and Debye temperature were evaluated. The hardness was measured by loading a diamond pyramid-type (with apex 136°) indenter into the surface of the specimen at room temperature, using the micro-Vickers hardness tester (MATSUZAWA SEIKI MHT-1).

The specific heat capacity was measured by the differential scanning calorimeter (DSC, ULVAC) apparatus, in the temperature range from room temperature to about 1200 K. The apparatus has a "triple-cell" system and an adiabatic temperature control system, which was originally developed by Takahashi et al. [21]. The principle of the apparatus is briefly summarized in the literature. The measurement was made in high purity argon (99.999 %) atmosphere with a flow rate of 100 ml/min. The accuracy of the apparatus was checked using an alpha-Al_2O_3 standard. The thermal diffusivity was measured by the laser flash method from room temperature to about 1200 K in vacuum (10^{-4} Pa) by using TC-7000 (ULVAC). The thermal diffusivity at a temperature obtained during the heating process also was checked during the cooling cycle. The thermal conductivity was calculated from the measured thermal diffusivity α, heat capacity C_P, and density ρ, using the following standard expression:

$$\lambda = \alpha C_P \rho . \quad (1)$$

RESULTS AND DISCUSSION

The conventional parameter describing the geometric distortion of the perovskite type compound is defined as a tolerance factor, t:

$$t = \frac{r_A + r_O}{\sqrt{2}(r_B + r_O)}, \quad (2)$$

where r_A, r_B, r_O are the ionic radius of each atom. Ordinarily, the value of "t" is within 0.75 to 1.1 for the perovskite type oxides. The cubic structure has a value near 1. As the value of "t" shifts from 1, the geometric distortion becomes large. The value of the tolerance factor "t" of $BaHfO_3$ and $SrHfO_3$ are shown in Table 1. The Shannon's values of the ionic radius [22] are used in the present study. Figure 1. shows the powder X-ray diffraction patterns at room temperature of the samples. It shows that the perovskite type single phase $BaHfO_3$ and $SrHfO_3$ are obtained in the present study. From the X-ray diffraction pattern, the crystal structures are evaluated as cubic perovskite with the lattice parameter of 0.4171 nm for $BaHfO_3$ and orthorhombic [10] perovskite with the lattice parameters of a= 0.5790 nm, b= 0.5851 nm, c= 0.8243 nm for $SrHfO_3$ at room temperature. The bulk densities of the samples are 91 % of the X-ray density. The chemical composition does not deviate from the stoichiometric composition. In addition, the impurities that originated during SPS technique were not confirmed. The sample characteristics are summarized in Table 1. $BaHfO_3$ and $SrHfO_3$ show white colors.

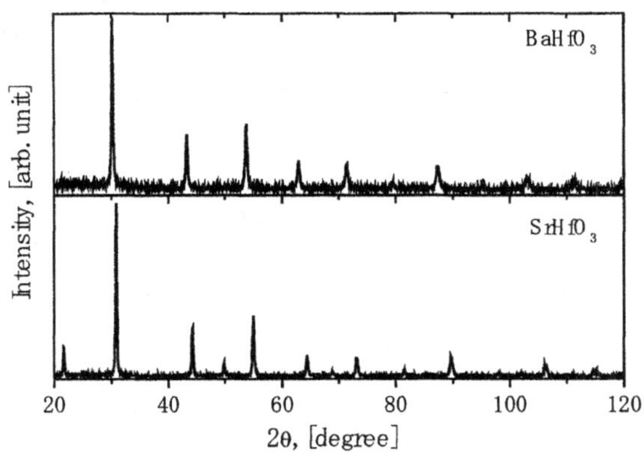

Figure 1. X-ray diffraction patterns of $BaHfO_3$ and $SrHfO_3$ at room temperature

Table 1. Sample characteristics and thermophysical properties of $BaHfO_3$ and $SrHfO_3$

			$BaHfO_3$	$SrHfO_3$
Crystal system at R.T.			Cubic	Orthorhombic
Lattice parameter at R.T.	[nm]	a	0.4171	0.5790
		b		0.5851
		c		0.8243
Relative density	[%T.D.]		91	91
Tolerance factor		t	1.016	0.958
Average linear thermal expansion coefficient	$[K^{-1}]$	α_l	7.91×10^{-6} (300-1500 K)	11.3×10^{-6} (300-1100 K)
Longitudinal wave velocity	[m/s]	V_L	5397	6118
Shear wave velocity	[m/s]	V_S	3245	3531
Young's modulus	[GPa]	E	194	220
Shear modulus	[GPa]	G	79.5	87.9
Compressibility	$[GPa^{-1}]$	β	8.78×10^{-3}	6.82×10^{-3}
Debye temperature	[K]	θ_D	438	490
Melting temperature	[K]	T_m	(2893)	3200
Vickers hardness	[GPa]	H_V	5.1	9.3
Heat capacity ($Cp=a+bT+cT^{-2}$)	$[JK^{-1}mol^{-1}]$	a	126	118
		b	4.99×10^{-3}	1.76×10^{-2}
		c	-2.01×10^{5}	-8.18×10^{5}
Thermal conductivity at R.T.	$[Wm^{-1}K^{-1}]$	λ	10.4	4.54

Figure 2 shows the temperature dependence of the lattice parameter of $BaHfO_3$ measured with the high temperature powder X-ray diffraction analysis. The thermal expansion coefficients was calculated from the measured lattice parameter, using the following equation:

$$\alpha_l = \frac{\partial a}{\partial T} \cdot \frac{1}{a_{300}}. \qquad (3)$$

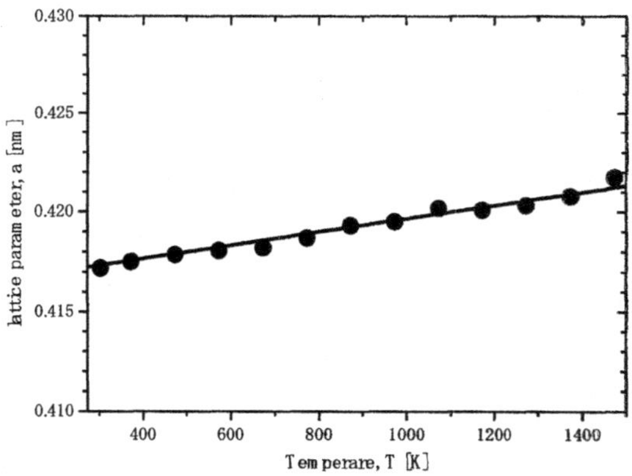

Figure 2. Temperature dependence of the lattice parameter for $BaHfO_3$

The average linear thermal expansion coefficients α_l are 7.91×10^{-6} K^{-1} for $BaHfO_3$ from 300 K to 1500 K and 11.3×10^{-6} K^{-1} for $SrHfO_3$ from 300 K to 1100 K. The melting temperature T_m of $BaHfO_3$ and $SrHfO_3$ is shown in Table 1. In the present study, the melting point of $BaHfO_3$ was not evaluated because of the unclear thermal arrest signal, so the reference value [8] is shown in Table 1. The values are 2893 K for $BaHfO_3$ and 3200 K for $SrHfO_3$. The melting temperature of $SrHfO_3$ is in good agreement with literature data [9]. The melting temperature of $SrHfO_3$ is higher than those of other perovskites, for example, the melting temperature of $SrTiO_3$ is 2170 K [23], and that of $SrZrO_3$ is 2883 K [20]. It is confirmed that the linear thermal expansion coefficient varies inversely as the melting temperature for many substances [24]. For the barium and strontium series perovskite type oxides, the products of α_l and T_m show approximately the same value (= 0.022) [11]. $BaHfO_3$ indicates similar tendency with those of other perovskites, but $SrHfO_3$ shows the different characteristics.

For an isotropic medium, the Young's modulus E, shear modulus G, compressibility β, and Debye temperature θ_D can be written in terms of the longitudinal sound velocity V_L and shear sound velocities V_S as follows [25,26]:

$$E = \frac{\rho V_S^2 (3V_L^2 - 4V_S^2)}{(V_L^2 - V_S^2)}, \tag{4}$$

$$G = \rho V_S^2, \tag{5}$$

$$\beta = \frac{1}{\rho\left(V_L^2 - \frac{4}{3}V_S^2\right)}, \tag{6}$$

$$\theta_D = \left(\frac{h}{k_B}\right) \cdot \left[\frac{9N}{4\pi V \cdot (V_L^{-3} + 2V_S^{-3})}\right]^{\frac{1}{3}}, \tag{7}$$

where h is the Plank constant, k_B is the Boltzmann constant, N is the number of atoms in a unit cell, and V is the unit cell volume. The elastic moduli and Debye temperature evaluated in the present study are summarized in Table 1. The Vickers hardness is shown in Table 1. In the present study, the applied load and loading time were chosen to be 1 kg (9.8 N) and 30 sec. The measurement was repeated 12 times for a sample, and the average value was calculated from the data in which the maximum and minimum values were excluded.

The temperature dependence of the heat capacity of $BaHfO_3$ and $SrHfO_3$ were determined by the scanning method. In the temperature range between 300 K and about 1300 K, the empirical equation for the C_P of $BaHfO_3$ is determined from the experimental data as follows:

$$C_P(JK^{-1}mol^{-1}) = 126 + 4.99 \times 10^{-3} T - 2.01 \times 10^5 / T^2. \tag{8}$$

On the other hand, the empirical equation for the C_P of $SrHfO_3$ is determined from the experimental values as follows:

$$C_P(JK^{-1}mol^{-1}) = 118 + 1.76 \times 10^{-2} T - 8.18 \times 10^5 / T^2. \tag{9}$$

The temperature dependence of the thermal conductivity of $BaHfO_3$ and $SrHfO_3$ is shown in Fig. 3. The thermal conductivity was corrected to 100 % of the theoretical density by using Schulz's equation [27]. The thermal conductivities of $BaHfO_3$ and $SrHfO_3$ at room temperature are 14.1 $Wm^{-1}K^{-1}$ and 5.20 $Wm^{-1}K^{-1}$, respectively. The thermal conductivities of $BaHfO_3$ and $SrHfO_3$ decrease with increasing temperature, which indicates that the phonon conduction is predominant. The thermal conductivities of $BaHfO_3$ and $SrHfO_3$ reach to the values, 5.05 $Wm^{-1}K^{-1}$ at 1275 K and 2.58 $Wm^{-1}K^{-1}$ at 864 K, respectively. These values are too high to utilize these materials for TBC application. The value of better TBC materials should have much lower thermal conductivity (<2 $Wm^{-1}K^{-1}$).

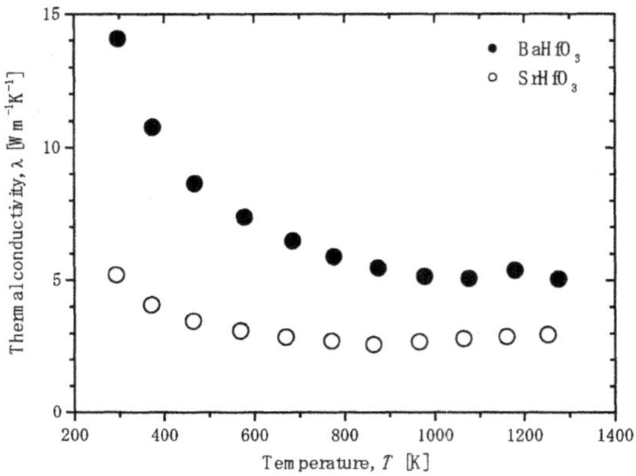

Fig. 3. Temperature dependence of the thermal conductivities for $BaHfO_3$ and $SrHfO_3$

CONCLUSION

The polycrystalline samples of alkaline earth perovskite type oxides, $BaHfO_3$ and $SrHfO_3$ were prepared. The thermophysical properties of $BaHfO_3$ and $SrHfO_3$ were measured and the relationships between the properties were studied. The average linear thermal expansion coefficients α_l are 7.91×10^{-6} K^{-1} for $BaHfO_3$ from 300 K to 1500 K and 11.3×10^{-6} K^{-1} for $SrHfO_3$ from 300 K to 1100 K. The melting temperatures are 2893 K for $BaHfO_3$ and 3200 K for $SrHfO_3$. It is confirmed that $SrHfO_3$ has very high melting temperature as well as high thermal expansion coefficient. The Young's moduli of $BaHfO_3$ and $SrHfO_3$ with 91 %T.D. are 194 and 220 GPa, respectively. The Vickers hardness of $BaHfO_3$ and $SrHfO_3$ are 5.1 and 9.3 GPa, respectively. In the temperature range between 300 and 1300 K, the empirical equation for the C_P of $BaHfO_3$ and $SrHfO_3$ is determined as follows:

$$C_P(JK^{-1}mol^{-1}) = 126 + 4.99\times10^{-3} \quad - 2.01\times10^{5}/ \quad ^{2}. \ (BaHfO_3)$$

$$C_P(JK^{-1}mol^{-1}) = 118 + 1.76\times10^{-2} \quad - 8.18\times10^{5}/ \quad ^{2}. \ (SrHfO_3)$$

The thermal conductivities of $BaHfO_3$ and $SrHfO_3$ at room temperature are 14.1 $Wm^{-1}K^{-1}$ and 5.20 $Wm^{-1}K^{-1}$, respectively.

REFERENCES

[1]J.G. Bednorz and K.A. Muller, "Strontium calcium titanate ($Sr_{1-x}Ca_xTiO_3$): an XY quantum ferroelectric with transition to randomness," *Phys. Rev. Lett.*, **52**, 2289-92 (1984).

[2]H.P.R. Frederikse, W.R. Thurber, and W.R. Hosler, "Electronic transport in strontium titanate,"

Phys. Rev., **134**, A442-5 (1964).
[3]C.S. Koonce, M.L. Cohen, J.F. Schooley, W.R. Hosler, and E.R. Pfeiffer, "Superconducting transition temperatures of semiconducting strontium titanate," *Phys. Rev.*, **163**, 380-90 (1967).
[4]V.E. Henrich, "The surfaces of metal oxides," *Rep. Prog. Phys.,* **48**, 1481-541 (1985).
[5]H. Muta, K. Kurosaki, and S. Yamanaka, "Thermoelectric properties of rare earth doped $SrTiO_3$," *J. Alloy. Comp.*, **350**, 292-5 (2003).
[6]K. Matsumoto, Y. Itoh, and T. Kameda, "EB-PVD process and thermal properties of hafnia-based thermal barrier coating," *Sci. Tech. Adv. Mater.*, **4**, 153-8 (2003).
[7]H.M. Wang, M.C. Simmonds, and J.M. Rodenburg," Manufacturing of YbAG coatings and crystallization of the pure and Li2O-doped Yb_2O_3-Al_2O_3 system by a modified sol-gel method," *Mater. Chem. Phys.*, **77**, 802-7 (2003).
[8]A.V. Shevchenko, L.M. Lopato, G.I. Gerasimyuk, and Z.A. Zaitseva, "Reactions in the systems hafnium dioxide - stronium oxides, hafnium dioxide-barium oxide and zirconium dioxide-barium oxide in the regions with a high content of a high content of hafnium dioxide (zirconium dioxide)," *Inst. Probl. Neorganicheskie Materialy*, **23**, 1495-9 (1987).
[9]M.P. Jorba, G. Tilloca, and R. Collongues, "Structure and properties of strontium zirconate; study of the phase diagram ZrO_2-SrO; extension of the investigation to other systems [of type] MO_2-M'O," *Intern. Symp. Magnetohydrodyn. Elec. Power Generation*, **3**, 1185-97 (1964).
[10]B.J. Kennedy, C.J. Howard, and B.C. Chakoumakos, "High-temperature phase transitions in $SrHfO_3$," *Phys. Rev. B*, **60**, 2972-5 (1999).
[11]S. Yamanaka, T. Maekawa, H. Muta, T Matsuda, S. Kobayashi, and K. Kurosaki, "Thermophysical properties of $SrHfO_3$ and $SrRuO_3$," *J. Solid State Chem.,* **177**, 3484-9 (2004).
[12]S. Yamanaka, M. Fujikane, T. Hamaguchi, H. Muta, T. Oyama, T. Matsuda, S. Kobayashi, and K. Kurosaki, "Thermophysical properties of $BaZrO_3$ and $BaCeO_3$," *J. Alloys and Comp.*, **359**, 109-13 (2003).
[13]S. Yamanaka, K. Kurosaki, T. Matsuda, and S. Kobayashi, "Thermal properties of SrCeO3," *J. Alloys and Comp.*, **352**, 52-6 (2003).
[14]T. Matsuda, S. Yamanaka, K. Kurosaki, and S. Kobayashi, "High temperature phase transitions of $SrZrO_3$," *J. Alloys and Comp.*, **351**, 43-6 (2003).
[15]S. Yamanaka, T. Matsuda, K. Kurosaki, and M. Uno, "Thermo-chemical and -physical properties of $BaMO_3$," *J. Nucl. Sci. Technol.*, **Supl.3**, 709-12 (2002).
[16]K. Kurosaki, T. Matsuda, M. Uno, S. Kobayashi, and S. Yamanaka. "Thermoelectric properties of $BaUO_3$," *J. Alloys and Comp.*, **319**, 271-5 (2001).
[17]S. Yamanaka, K. Kurosaki, T. Matsuda, and M. Uno, "Thermophysical properties of $BaUO_3$," *J. Nucl. Mater.*, **294**, 99-103 (2001).
[18]H. Muta, K. Kurosaki, and S. Yamanaka, "Thermoelectric properties of doped $BaTiO_3$–$SrTiO_3$ solid solution," *J. Alloys and Comp.*, **368**, 22-4 (2004).
[19]K. Kurosaki, T. Oyama, H. Muta, M. Uno and S. Yamanaka, "Thermoelectric properties of perovskite type barium molybdate," *J. Alloys and Comp.*, **372**, 65-9 (2004).
[20]S. Yamanaka, K. Kurosaki, T. Matsuda, T. Oyama, H. Muta, S.Kobayashi, and M. Uno, "Thermophysical properties of perovskite type strontium cerate and zirconate" *J. Am. Ceram. Soc.*, **88**, 1496-9 (2005).
[21]Y. Takahashi and M. Asou, "Development of a high-temperature differential scanning calorimeter equipped with a triple-cell system," *Thermochimica Acta*, **223**, 7-22 (1993).
[22]R.D. Shannon, "Revised effective ionic radii and systematic studies of interatomic distances in halides and chalcogenides," *Acta Cryst.*, **A32**, 751-67 (1976).

[23]N. Sata, H. Matsuta, Y. Akiyama, Y. Chiba, S. Shin, and M. Ishigame, "Fabrication of proton conducting thin films of $SrZrO_3$ and $SrCeO_3$ and their fundamental characterization," *Solid State Ionics*, **97**, 437-441 (1997).
[24]L.G. Van Uitert, H.M. O'Bryan, M.E. Lines, H.J. Guggenheim, and G. Zydzik, "Thermal expansion - an empirical correlation," *Mat. Res. Bull.*, **12**, 261-8 (1977).
[25]K. Yamada, S. Yamanaka, and M. Katsura," *Tech. Rep. Osaka Univ.*, **47**, 181 (1997).
[26]M. Fukuhara and I. Yamauchi, "Temperature dependence of the elastic moduli, dilational and shear internal frictions and acoustic wave velocity for alumina, (Y)TZP and β'-Sialon ceramics," *J. Mater. Sci.*, **28**, 4681-8 (1993).
[27]B. Schulz, "Thermal conductivity of porous and highly porous materials," *High Temp.-High Press.*, **13**, 649-60 (1981).

THERMOPHYSICAL PROPERTIES OF SINTERED SrY_2O_4 AND THE RELATED COMPOUNDS APPLICABLE TO THERMAL BARRIER COATING MATERIALS

Authors: Takanori Tanaka, Ken Kurosaki, Takuji Maekawa, Hiroaki Muta, Masayoshi Uno and Shinsuke Yamanaka
Affiliation: Department of Nuclear Engineering, Graduate School of Engineering, Osaka University
Address: 2-1 Yamadaoka
City, State, Zip: Suita, Osaka, 565-0871, Japan

ABSTRACT

Polycrystalline samples of SrY_2O_4 were prepared by reacting the appropriate amounts of Y_2O_3 and $SrCO_3$. The sample of SrY_2O_4 was sintered by using Spark Plasma Sintering (SPS). The dense sample with about 99 % of theoretical density is obtained in the present study. The crystal structure and lattice parameters were evaluated with an X-ray diffraction method. SrY_2O_4 have an orthorhombic structure with the space group Pnam. The thermal expansion coefficient was measured with a differential dilatometer in the temperature range from 300 K to 1773 K. The average liner thermal expansion coefficient is $10.7\times10^{-6}\ K^{-1}$. The heat capacity was measured using a differential scanning calorimeter (DSC). The thermal diffusivity was measured with a laser flash method in the temperature range between room temperature and about 1273 K. The thermal conductivity was evaluated from the measured heat capacity and thermal diffusivity. The elastic moduli and Debye temperature were evaluated from the longitudinal and shear sound velocities, which were measured with an ultrasonic pulse-echo method at room temperature in air. The Young's modulus of SrY_2O_4 is 162 GPa.

INTRODUCTION

Thermal barrier coatings (TBCs) of partly Y_2O_3 stabilized ZrO_2 (PYSZ) are widely used to protect hot section parts of aircraft and land-based turbines by reducing the temperature of metal substrates. Their continued development is essential for improving the efficiency and performance of gas turbines by reducing the temperature of metal substrates. However this standard material has a limited temperature capability due to accelerated sintering and phase transitions at high temperatures. In general, TBC materials have to fulfill most of the following requirements, such as a stable phase, low thermal conductivity, high thermal expansion coefficient, chemical resistance, low sintering rate, and high fracture toughness. A worldwide effort has been undertaken to identify new candidates for a TBC application.[1, 2]

Against this background, we focus on SrY_2O_4 and BaY_2O_4 as the new TBC materials. The compounds are a 1:1 compound in the SrO-Y_2O_3 and BaO-Y_2O_3 pseudo binary systems. The crystal structure is the orthorhombic $CaFe_2O_4$ structure with the space group Pnam as shown in Fig. 1.[3-7] Although the crystal structure, high temperature stability, and optical properties of SrY_2O_4 have reported already, the thermal and mechanical properties are scarcely reported.[4-5, 7-11] In the present study, polycrystalline sintered high-density samples were prepared and the thermal and mechanical properties were measured.

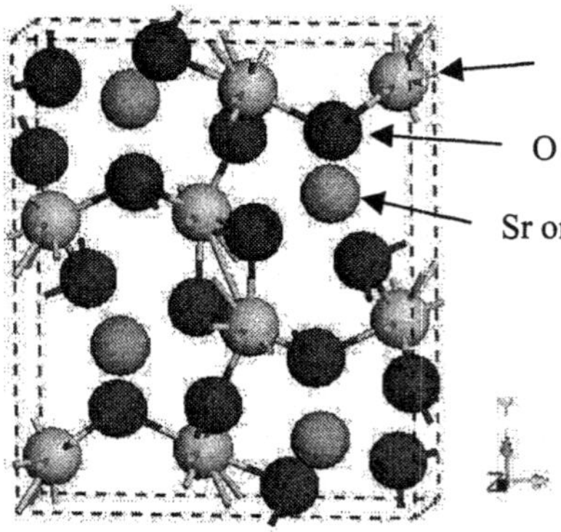

Fig. 1. Crystal structure of SrY_2O_4 and BaY_2O_4.

EXPERIMENTAL

The polycrystalline samples of SrY_2O_4 and BaY_2O_4 were prepared by mixing the appropriate amounts of Y_2O_3 (Furuuchi Chemical, 99.9 %) and $SrCO_3$ (Aldrich Chemical, 99.9+ %) or $BaCO_3$ (Furuuchi Chemical, 99.9 %) followed by reacting at 1273 K for 10 h 2 times in air. After that, the SrY_2O_4 powder was placed into a 20 mm diameter graphite die and given a Spark Plasma Sintering (SPS, SUMITOMO COAL MINING Dr Sinter SPS-1020 apparatus) at 1773 K for 3 minutes under argon atmosphere.

The powder X-ray diffraction (RIGAKU RINT2000) was performed at room temperature using Cu-Kα radiation. In order to determine the chemical composition, the SEM-EDX analysis was performed using HITACHI S-2600H SEM instrument equipped with the energy-dispersive HORIBA EX-200 system. The oxygen analysis was performed using HORIBA EMGA-550 to identify the oxygen concentration of the samples. The bulk density of the sample was determined by the geometric measurement.

The thermal expansion characteristics were measured by a differential dilatometer in the temperature range 300 to 1773 K in air. The melting temperature was measured by the thermal arrest method under a reducing atmosphere.

The thermal diffusivity was measured by the laser flash method from room temperature to about 1200 K in vacuum (10^{-4} Pa) by using TC-7000 apparatus (ULVAC). The thermal diffusivity at a temperature obtained during the heating process also was checked during the cooling cycle. The thermal conductivity λ was calculated from the thermal diffusivity D_{th}, heat capacity C_P, and density ρ, using the following standard expression:

$$\lambda = D_{th} C_P \rho. \quad (1)$$

The specific heat capacity was measured with the differential scanning calorimeter (DSC, ULVAC) apparatus, in the temperature range from room temperature to about 1200 K. The apparatus has a "triple-cell" system and an adiabatic temperature control system, which was originally developed by Takahashi et al.[12] The principle of the apparatus is briefly summarized in the literature. The measurement was made out in high purity argon (99.999 %) atmosphere with the flow rate of 100 ml/min. The accuracy of the apparatus was checked using an alpha-Al_2O_3 standard.

The longitudinal and shear sound velocities were measured by the ultrasonic pulse-echo method (NIHON MATECH Echometer 1062) at room temperature in air. The sintered sample was bonded to a 5 MHz longitudinal or shear sound wave echogenic transducer. From the sound velocities, the elastic moduli and Debye temperature were evaluated.

RESULTS and DISCUSSION

The powder X-ray diffraction patterns at room temperature of the samples are shown in Fig. 2, together with the literature data.[5, 6] From this figure, it is observed that the phase of SrY_2O_4 and BaY_2O_4 is formed in the present study. Small peaks corresponding to Y_2O_3 are observed in this figure. The lattice parameters evaluated from the X-ray diffraction patterns are summarized in Table 1. The values of the lattice parameters agree with the literature data.[5]

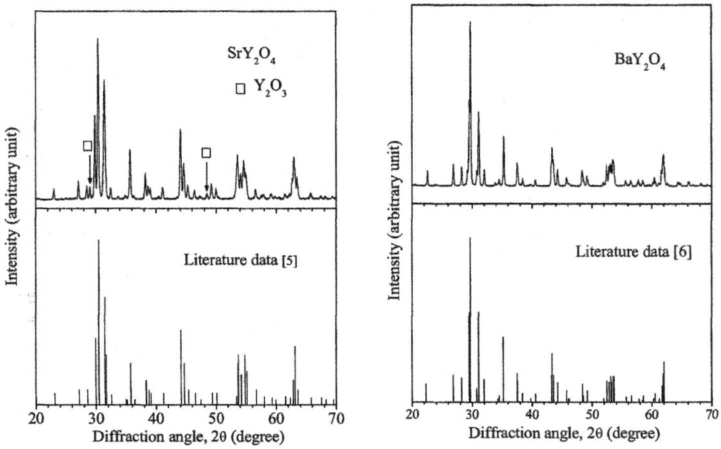

Fig. 2. X-ray diffraction patterns of SrY_2O_4 and BaY_2O_4, together with the literature data.[5, 6]

Table 1. Structure characteristics of SrY_2O_4 and BaY_2O_4

			SrY_2O_4	BaY_2O_4
Crystal system			Orthorhombic	
Space group			Pnam	
Lattice parameters at room temperature	a	(nm)	1.0077	1.0364
	b	(nm)	1.1914	1.2081
	c	(nm)	0.3410	0.3441

Only for SrY_2O_4, the high density bulk sample was obtained by the SPS process. The bulk density of the sample is 98.8 % of the theoretical density (5.28 gcm^{-3}). The chemical composition determined by the EDX and oxygen chemical analyses does not deviate from the stoichiometric composition. The obtained sample shows white color.

Figure 3 shows the thermal expansion behavior of SrY_2O_4. The size of the sample increases with temperature, indicating a positive thermal expansion in Fig. 3 (a). The linear thermal expansion coefficient increases with temperature as shown in Fig. 3 (b). The average linear thermal expansion coefficient in the temperature range between 300 and 1773 K is 10.7×10^{-6} K, which is very consistent with the literature data obtained from the High temperature X-ray diffraction analysis. This value is not enough large to utilize this material for TBC application.

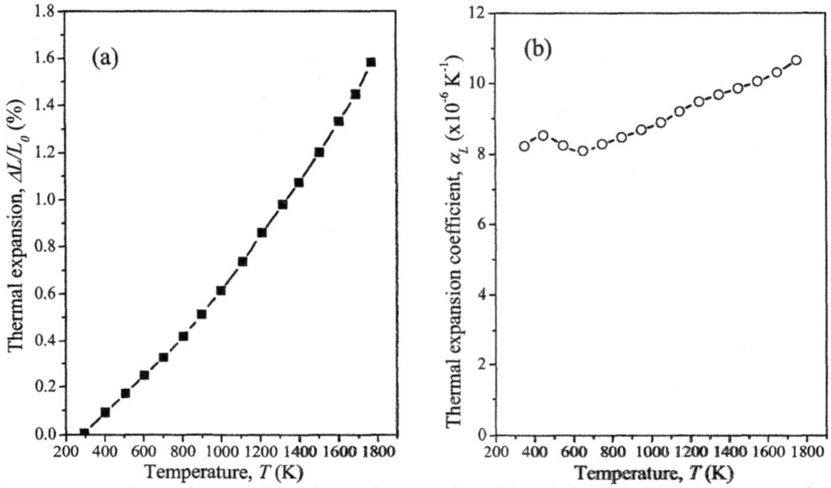

Fig. 3. Temperature dependence of the thermal expansion (a) and the linear thermal expansion coefficient (b) from 300 to 1773 K for SrY_2O_4.

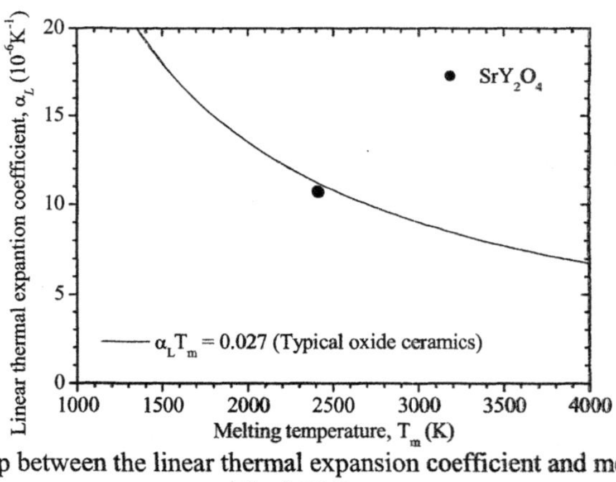

Fig. 4. Relationship between the linear thermal expansion coefficient and melting temperature for SrY_2O_4.

The melting temperature of SrY_2O_4 determined by the thermal arrest method is 2413 K, which is slightly lower than that in the literature.[3] The relationship between the liner thermal expansion coefficient and melting temperature of SrY_2O_4 is shown in Fig. 4. In this figure, the data of other substances are plotted for comparisons. From this figure, it is confirmed that SrY_2O_4 has the moderate high thermal expansion coefficient as well as high melting temperature, compared with those of the typical oxide ceramics. This characteristic is desirable for TBC application.

For an isotropic medium, the Young's modulus E, shear modulus G, compressibility β, and Debye temperature θ_D can be written in terms of the longitudinal sound velocity V_L and shear sound velocities V_S as follows.

$$E = \frac{\rho V_S^2 (3V_L^2 - 4V_S^2)}{(V_L^2 - V_S^2)}, \quad (7)$$

$$\beta = \left(V_L^2 - \frac{4}{3} V_S^2 \right)^{-1}, \quad (8)$$

$$\theta_D = \left(\frac{h}{k_B} \right) \cdot \left[\frac{9N}{4\pi V \cdot (V_L^{-3} + 2V_S^{-3})} \right]^{\frac{1}{3}}, \quad (9)$$

where ρ is the sample density, h is the Plank constant, k_B is the Boltzmann constant, N is the number of atoms in a unit cell, and V is the unit cell volume. The elastic moduli and Dybye temperature of SrY_2O_4 evaluated in the present study are summarized in Table 2.

Table 2. Thermophysical properties of SrY_2O_4

Density	ρ	gcm^{-3}	5.28
		%T.D.	98.8
Melting temperature	T_m	(K)	2413
Young's modulus	E	(GPa)	162
Compressibility	β	(GPa^{-1})	8.61×10^{-3}
Debye temperature	θ_D	(K)	471
Thermal conductivity (300 K)	λ_{300}	($Wm^{-1}K^{-1}$)	7.6
(1200 K)	λ_{1200}	($Wm^{-1}K^{-1}$)	3.4

The temperature dependence of the heat capacity of SrY_2O_4 is obtained by the scanning method using the triple-cell DSC. The peaks from the phase transitions are not observed in the C_P curve of SrY_2O_4. In the temperature range between 300 and 1273 K, the empirical equation for the C_P of SrY_2O_4 is determined by the experimental data as follows:

$$C_P = 186.6 - 0.0312T + 3.00\times10^{-5}\ ^{2} - 2.51\times10^{-6}\ ^{-2}\ (JK^{-1}mol^{-1}). \quad (10)$$

The thermal diffusivity was measured by the laser flash method. The thermal conductivity of SrY_2O_4 was evaluated from the thermal diffusivity, heat capacity and density. The temperature dependences of the thermal diffusivity and conductivity of SrY_2O_4 are shown in Fig. 5 (a) and (b), respectively.

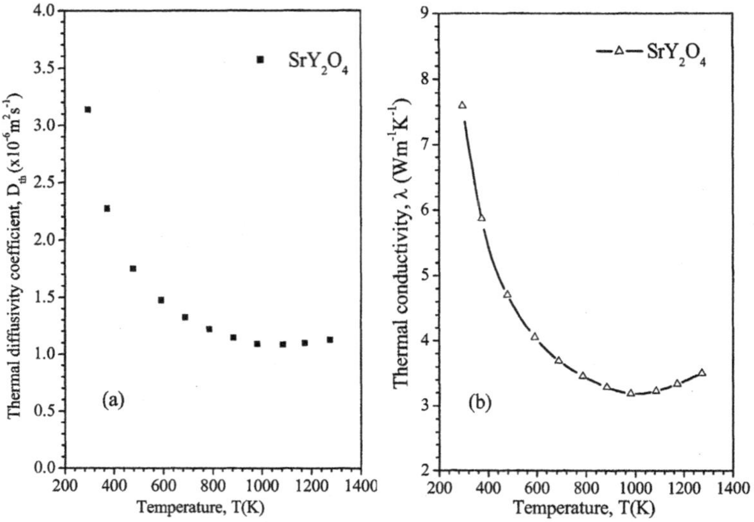

Fig. 6 Temperature dependence of the thermal diffusivity of SrY_2O_4 (a) and thermal conductivities of SrY_2O_4 (b).

Both the thermal diffusivity and conductivity decrease with increasing temperature, which indicates that the phonon conduction is predominant. The thermal conductivities at 300 K and 1200 K are 7.6 $Wm^{-1}K^{-1}$ and 3.4 $Wm^{-1}K^{-1}$, respectively.

SUMMARY

The polycrystalline samples of SrY_2O_4 and BaY_2O_4 were prepared by the solid state reactions. The crystallographic properties were studied by the XRD analysis. The high density sintered sample of SrY_2O_4 was prepared by the SPS, and the thermophysical properties were measured. The crystal structure is orthorhombic system with the space group Pnam. The average linear thermal expansion coefficient is 10.7×10^{-6} K^{-1} in the temperature range from 300 to 1773 K for SrY_2O_4. The Young's modulus and Debye temperature of SrY_2O_4 are 162 GPa and 471 K, respectively. The thermal conductivity of SrY_2O_4 decreases with increasing temperature. The values at 300 K and 1200 K are 7.6 $Wm^{-1}K^{-1}$ and 3.4 $Wm^{-1}K^{-1}$, respectively.

REFERENCES

[1]K. Matsumoto, Y. Itoh, T Kameda, "EB-PVD process and thermal properties of hafnia-based thermal barrier coating," *Sci. Tech. Adv. Mater.*, **4**, 153-158 (2003).

[2]H.M. Wang, M.C. Simmonds, J.M. Rodenburg, "Manufacturing of YbAG coatings and crystallisation of the pure and Li_2O-doped Yb_2O_3–Al_2O_3 system by a modified sol–gel method," *Mater. Chem. Phys.*, **77**, 802-807 (2003).

[3]S.G. Tresvyatskii, L.M. Lopato, A.E. Kushchevskii, A.V. Shevchenko, "Phase diagrams of yttrium sesquioxide-strontium oxide and ytterbium sesquioxide-strontium oxide systems." *Neorg. Mater.*, **7**, 1808-1811 (1971).

[4]Hk. Muller-Buschbaum, "Strontium yttrium oxide, SrY_2O_4," *Z. Anorg. Allg. Chem.*, **358**, 138-146 (1968).

[5]JCPDS 32-1272.

[6]JCPDS 27-44.

[7]J. G. Pepin, "Crystal data for $SrRE_2O_4$ (RE = rare earth + Y, In)," *J. Appl. Cryst.*, **14**, 70-71 (1981).

[8]S. Arul Antony, K. S. Nagaraja, G. L. N. Reddy, and O. M. Sreedharan, "A polymeric gelcum auto-combustion method for the lower temperature synthesis of SrR_2O_4 (R = Y, La, Sm, Eu, Gd, Er or Yb)," *Mater. Lett*, **51**, 414-419 (2001).

[9]S. Arul Antony, R. Pankajavalli, K. S. Nagaraja and O. M. Sreedharan, "High temperature stability of SrY_2O_4 by SrF_2-based emf. method," *Mater. Lett.*, **57**, 469-474 (2002).

[10]W. Xu, W. Jia, I. Revira, K. Monge, and H. Liu, "Optical properties of multiple sites of Eu^{3+} in SrY_2O_4 single-crystal fibers," *J. Electrochem. Soc.*, **148**, H176-H178 (2001).

[11]S.J. Park, C.H. Park, B.Y. Yu, H.S. Bae, C.H. Kim, and C.H. Pyun, "Structure and luminescence of SrY_2O_4:Eu," *J. Electrochem. Soc.*, **146**, 3903-3906 (1999).

[12]Y. Takahashi and M. Ashou, "Development of a high-temperature differential scanning calorimeter equipped with a triple-cell system," *Thermochem. Acta.*, **223**, 7-22 (1993).

ELECTRICAL PROPERTIES OF MICROWAVE PLASMA CHEMICAL VAPOR DEPOSITED DIAMOND THIN FILMS

R. Ramamurti, R. N. Singh
Department of Chemical and Materials Engineering,
University of Cincinnati,
Cincinnati, OH 45221-0012

P. B. Kosel
Department of Electrical and Computer Engineering and Computer Science,
University of Cincinnati,
Cincinnati, OH 45221-0077

ABSTRACT

Diamond films are prepared by microwave plasma enhanced chemical vapor deposition (MPECVD) on silicon (100) substrates using Ar-H_2-CH_4 plasmas. The current-voltage characteristics of undoped microcrystalline diamond films are measured as functions of temperature and applied voltage. Currently these films have tremendous applications in high temperature electronics. Variation of electrical properties of these films like resistivity, activation energy and carrier concentrations are studied with temperature and applied voltage. This can be related with the types of conduction taking place in different regimes.

Keywords: CVD, microwave plasma, diamond thin films, I-V characteristics, electrical resistivity, carrier concentrations, activation energy, conduction mechanism

INTRODUCTION

The unique combination of physical and electrical properties of diamond such as large band gap, high resistivity, high thermal conductivity, large carrier mobility, low dielectric constant, and good transmittance from the UV to IR makes it suitable for high temperature, high power and high speed electronic devices. The electrical properties of the chemical vapor deposited (CVD) diamond films depend not only on the properties of the bulk, but also on the device preparation techniques such as surface treatments of the film and contact fabrication [1-3]. Consequently, measurement of the electrical properties of the polycrystalline CVD diamond film is required for fabrication of the electronic devices.

The resistivity of natural diamond is approximately constant (10^{15}-10^{16} Ω.cm) between room temperature and 200 °C [4]. Natural diamond does not contain any hydrogen and this is the reason for the high resistivity. In comparison, CVD diamond has resistivities on the order of $\sim 10^{12}$ Ω.cm. This may be due to the hydrogen passivation of the deep traps. It is shown that the measured electrical characteristics of chemical vapor deposited (CVD) diamond can be superior to those of natural diamond if the diamond films are devoid of impurities or defects [4]. In addition, the conductivity of CVD diamond films decreased by several orders of magnitude by thermal annealing, and the subsequent treatment with hydrogen plasma restored the conductivity to the original value [5].

Vandersande et. al. [4] measured resistivity of CVD diamond films at high temperatures. These films were grown by Raytheon, Crystallume and Norton companies and showed a resistivity of $\sim 10^{16}$ Ω.cm at room temperature, which is one order of magnitude higher than that of natural diamond. This indicates that the quality of CVD diamond films were good. The samples resistivities decreased from 10^{16} Ω.cm to 10^5 Ω.cm over a temperature range of 25-1000 °C.

The electrical properties of diamond films can also change with grain size. The significant variation of resistivity with grain size has been attributed to either hydrogen passivation of defects or the presence of disordered graphitic regions between grains in the as-grown films [6]. The decrease in grain size could result in decreasing mean carrier path length [7]. The graphitic regions on the surface can be chemically etched in a saturated solution of CrO_3 and H_2SO_4 at 170 °C. This increases the resistivity by a few orders of magnitude [8, 9].

Typical activation energy values for CVD diamond is of in the range 1.5-2 eV. Many authors have attributed this to be due to nitrogen-related trace impurities (~0.01 ppm) [10].

In this work we are interested in determining the variation of electrical properties like resistivity, activation energy and carrier concentration of undoped microcrystalline diamond films.

EXPERIMENTAL PROCEDURES

An electron cyclotron resonance microwave plasma enhanced chemical vapor deposition (ECR-MPECVD) reactor was used to deposit polycrystalline diamond films on silicon substrate. More details on processing can be obtained from our previous publications [11-13]. Silicon wafers, p-type (100) diced into squares (25 x 25 x 0.5 mm) was used as substrates for deposition of PCD. The substrate activation procedure included ultrasonic treatment with a slurry of 20-40 μm diamond grit in 150 ml ethanol for two hours. This process was followed by rinsing in acetone, ethanol, water and drying in a N_2 jet before placing into the reactor. Prior to deposition, a 30 minute plasma-etching was done in the hydrogen plasma to clean the Si substrate off the native oxide layer. Mixtures of Ar, H_2, and CH_4 were used as the reactant gases during deposition. An optimized set of deposition parameters for these experiments include are shown in Table 1.

Table 1. Deposition Conditions for microcrystalline CVD diamond films

Sample ID	T_s (°C)	Pressure (torr)	Power (watts)	H_2 (sccm)	Ar (sccm)	CH_4 (sccm)	Thickness (μm)	Grain size (nm)
100703 ArLT	370	95	900	39	60	1	11.45	1600

The diamond films were characterized by Raman spectroscopy. The Raman system utilized a model T 64000, Jobin Yvon triple monochromator system, equipped with an optical multi-channel detector-CCD array with Olympus BX-41 microscope attachment. The Raman analysis of diamond films was done using Ar^+ ion laser at a wavelength of 514.5 nm. The 6-10 mW power was focused on to a spot size of 2 μm diameter on the film and the spectra were accumulated at room temperature for 60 seconds. A scanning electron microscope (Hitachi S4000) was used to examine the surface morphology of the samples. The electrical properties of PCD films were characterized by I-V measurements in which test capacitors were created. The details of this technique have been explained in our previous publications [13]. I-V measurements were made on PCD films using this set-up as shown in Fig. 1b.

A special rig was fabricated to measure the I-V characteristics over a range of temperatures. It consisted of a small enclosure, which was evacuated using a rotary pump to a pressure of ~10^{-3} torr (measured using a Hastings vacuum gauge SV-1S). The sample was heated by a 600 W tungsten lamp to the desired temperature. An Omega CN76000 temperature controller maintained the temperatures up to 500 °C. A Boonton 72-4B capacitance meter was used to measure the capacitance of the free-standing diamond films over a frequency range of 1 kHz to 1 MHz. An HP 6522A power supply used to generate a voltage up to 2000 volts. The voltage was read using a Keithley 177 microvolt digital multimeter. The current is measured by a Keithley 410A pico ammeter. A thin gold wire was used as a contact probe tip on the dot electrodes. The measurements were done after obtaining a pressure of ~10^{-3} torr in the measurement region using the rotary pump. A voltage was applied across the electrodes at room temperature and current was measured using an ammeter. Subsequently, the voltage was increased in steps of 50 V to 200 V and the corresponding current reading was obtained. The temperature was then increased to 100 °C and then in steps of 50 °C to 200 °C measurements were made at each temperature.

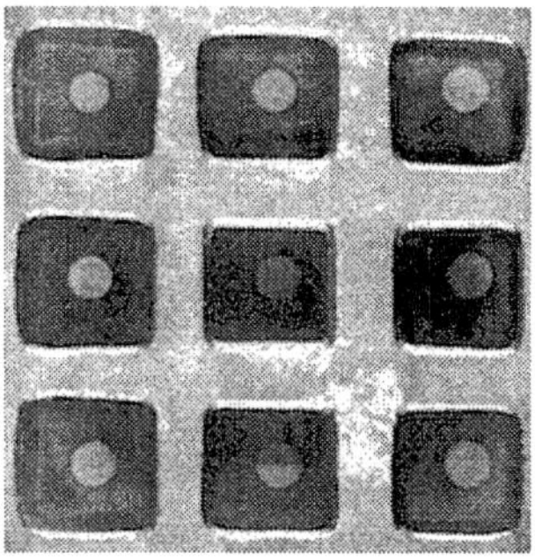

(a)

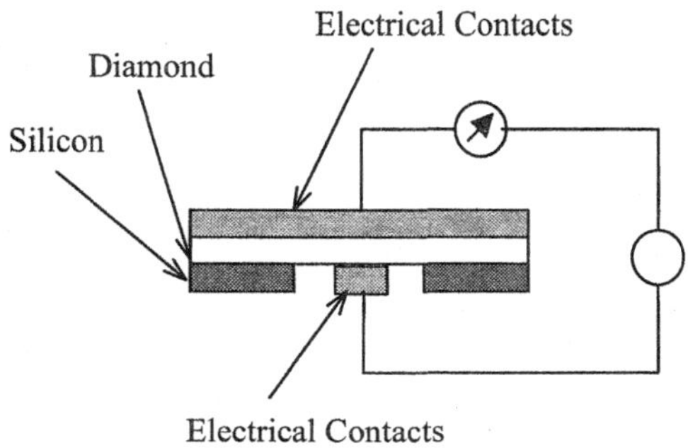

(b)

Fig. 1. (a) PCD DOT capacitor test structures fabricated on free-standing 10 μm thick PCD film, (b) Schematic of I-V measurement set-up on free-standing CVD diamond

RESULTS AND DISCUSSION

Scanning electron microscopy image of a microcrystalline diamond (MCD) film is shown in Fig. 2a. The grain size of MCD is 1600 nm. It can be seen that the grains are faceted with very little secondary nucleation at the grain boundaries.

The Raman spectroscopy result of MCD is shown in Fig. 3. The intensity of the Raman peak with center 1332.5 cm^{-1} for MCD is large and its width small indicating that the quality of MCD in terms of sp^3 content is good. Raman scattering intensity in the 1400-1600 cm^{-1} region is not very well pronounced. The development of the "graphite" band is due to π-bonded carbon at the grain boundaries [14].

Fig. 2. SEM images of microcrystalline diamond (Sample ID 100703ArLT)

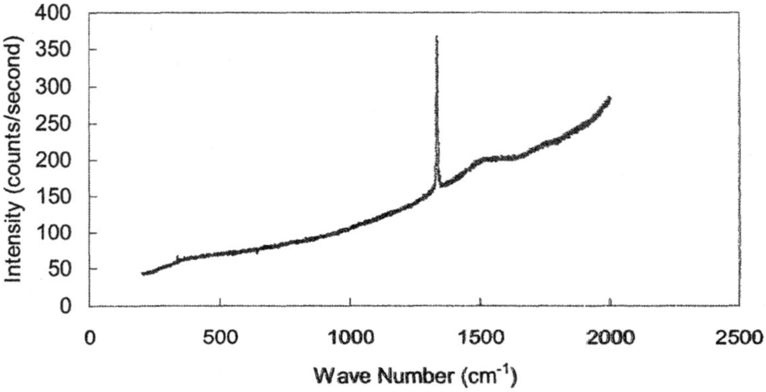

Fig. 3. Raman spectra of microcrystalline diamond (Sample ID 100703ArLT)

Electrical properties of microcrystalline diamond films are done over the voltage range of 0-200 volts and the current range of nanoamps to milliamps. The resistivity, carrier concentration, and activation energies were obtained for these samples in the temperature range of 25-300 °C.

Current (I)-voltage (V) and resistivity-voltage data for sample 100703ArLT (grain size=1600 nm) are shown in Fig. 4. The I-V data shown in Fig. 4a are obtained between 25 °C to 300 °C. These results show high-resistivity diamond films up to 150 °C where the current is in the range of 10^{-9} A over the voltage range of 50-200 V. At 200 °C, current increases from 2.6×10^{-9} A to 2.2×10^{-7} A. At 250-300 °C, the current increased from $\sim10^{-8}$ to $\sim10^{-6}$ A as the applied voltage is changed from 50 to 200 V. The I-V response is linear at lower voltages (up to 100 V) and non-linear at higher voltages (100-200 V). The non-linear response is apparently more prominent at 200 °C and above as seen in Fig. 4a. The data fits to an exponential function best since the R^2 fitting parameter values were in the range of 0.98 to 0.99. The data has good reproducibility with an error of 2%. This fit governs the Hill's conduction mechanism. The change in resistivity with sample temperature is shown in Fig. 4b. These results indicate that the resistivity is almost constant at $\sim10^{11}$ Ω.cm over the temperature range and voltage range of 25-

150 °C and 50-200 V respectively. At 200 °C, 250 °C and 300 °C the resistivity is ~10^{11}, 10^{10} and 10^{9} Ω.cm respectively at 50 volts and these values decrease by one order of magnitude at 200 V.

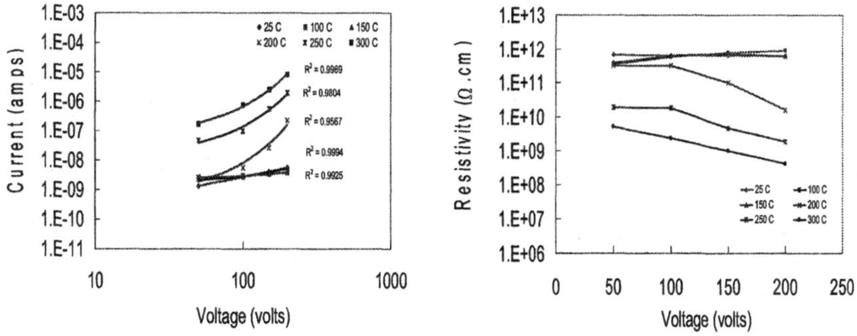

(a) (b)

Fig. 4. (a) Current-Voltage characteristics, (b) Resistivity versus Voltage measurements for Device #2 of Sample 100703ArLT (grain size=1600 nm)

Intrinsic conductivity is dominant at high temperatures when the number of intrinsic carriers in the conduction band is dominant due to thermal excitation from the valence band. The slope of the linear regime at high temperatures can be used to determine the activation energy for the intrinsic conduction. The intrinsic conductivity of a semiconductor can be expressed as

$$\sigma = n_i e (\mu_e + \mu_h) \tag{1}$$

The intrinsic carrier concentration (n_i) of the semiconductor is given as

$$n_i = p_i = 2(2\pi kT/h^2)^{3/2}(m_e m_h)^{3/4} e^{-(Ea/2kT)} \tag{2}$$

Where, σ = conductivity, n_i = intrinsic electron concentration, n_h = intrinsic hole concentration, μ_e = electron mobility, μ_h = hole mobility, m_e = electron mass, m_h = hole mass, T = sample temperature, E_a = activation energy for conduction, k=boltzmann's constant

Conventional theory [15] suggests that the mobility of the semiconductor varies as $\mu \sim T^{-1.5}$ for a lattice scattering mechanism and $\mu \sim T^{1.5}$ for an ionized impurity scattering mechanism.

From Eqs. (1) and (2) we get,

$$\sigma = \sigma_o e^{-(Ea/2kT)} \tag{3}$$

and

$$\ln \sigma = \ln \sigma_o - E_a/2kT \tag{4}$$

where, $\sigma_o = 2e(2\pi k/h^2)^{3/2}(m_e m_h)^{3/4}$

The intrinsic carrier concentration can be calculated for each voltage using the different temperatures and activation energies. The activation energy can be determined from the intrinsic regime (high temperature limit) of ln σ vs. 1/T plot.

The variation of ln σ with 1/T of sample 100703ArLT is shown in Fig. 5a. The activation energy determined from the slope of the high temperature (intrinsic regime) is shown in Fig. 5b.

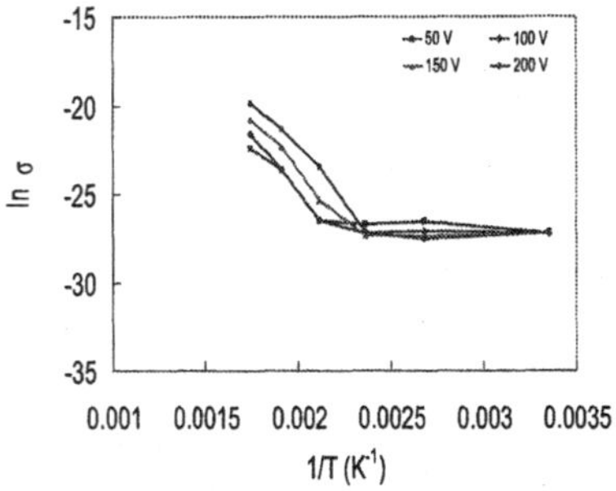

(a)

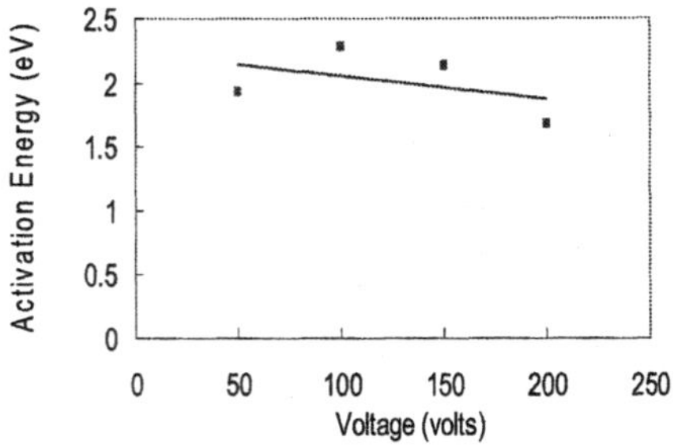

(b)

Fig. 5. Variation of (a) ln σ with 1/T and (b) Activation energy in the intrinsic regime with applied voltage for device #2 of sample 100703ArLT (grain size = 1600 nm)

At 50 V the activation energy is 2.14 eV and decreases to 1.87 eV at 200 V. These values suggest that there could be hydrogen trap levels just below the Fermi level. As hydrogen leaves the trap sites, carrier transport becomes more difficult thus leading to higher activation energy for conduction [16]. These low activation energies might be due to Fermi levels located in the band tail (at 0.75 – 1 eV above the valence band) [17] leading to variable range hopping (VRH) conduction described by Mott [18]. Some authors have attributed this to be due to the presence of graphitic phases in the grain boundaries. Thus the conduction is based on activation at shallow levels [8]. The intrinsic carrier concentration calculated using Eq. (2) is given in Fig. 6. It increases from 4.5×10^{7} cm^{-3} at 25 °C to 4.5×10^{14} cm^{-3} at 200 °C at 50 V. At 200 volts these values are 8.7×10^{9} cm^{-3} and 2.2×10^{16} cm^{-3} at 25 °C and 200 °C, respectively. This indicates that applied voltage decreases the activation barrier for conduction. A Hill's conduction mechanism has been established for the MCD film based on the fit for the current-voltage data. This has not been discussed in detail.

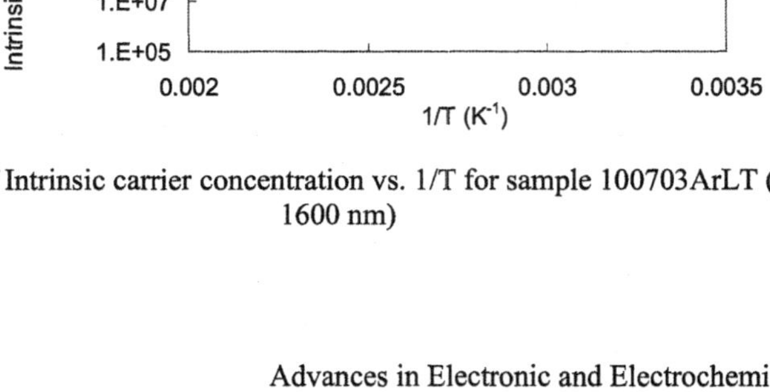

Fig. 6. Plot of Intrinsic carrier concentration vs. 1/T for sample 100703ArLT (grain size = 1600 nm)

CONCLUSIONS

A technique for fast fabrication of arrays of parallel-plate (DOT) capacitors on thin freestanding PCD films was used to prepare samples for electrical characterization. Electrical properties of microcrystalline (MCD) films are determined from I-V measurements. The larger grain size MCD films have very good diamond quality. The resistivity of the microcrystalline diamond film at an applied voltage of 50 V is of the order of 10^{11} Ω.cm at 25 °C and 10^{9} Ω.cm at 300 °C. The activation energy for conduction decreased with increasing voltage. The activation energies of microcrystalline diamond ranges from 2.14 eV to 1.87 eV in the voltage range 100-200 V. Electrically active traps above the valence band may be present that can contribute to the low activation energies. The intrinsic carrier concentrations for microcrystalline diamond varied from $\sim 10^{7}$ cm^{-3} at 25 °C to $\sim 10^{14}$ cm^{-3} at 200 °C at 50 V. These values increased to $\sim 10^{9}$ cm^{-3} and $\sim 10^{16}$ cm^{-3} at 200 V. The Hill's conduction mechanism fits the current-voltage data best. Further discussion needs to be done in order to explain this.

ACKNOWLEDGMENTS

The authors would like to thank Mr. Mahmood Samiee for the electrical measurements, Uday and Vamsi Krishna for Raman measurements. This project was supported by National Science Foundation under Grant Nos. ECS-0070004 and CMS-0210351. Any opinions, findings, and conclusions or recommendations expressed in this material are those of the authors and do not necessarily reflect the views of the National Science Foundation.

REFERENCES

[1]K. Das, V. Venkatesan, K. Miyata, D. L. Dreifus, J. T. Glass, "A Review of the Electrical Characteristics of Metal Contacts on Diamond", Thin Solid Films 212, 19-24 (1992).

[2]S. A. Grot, G. S. Gildenblat, C. W. Hatfield, C. R. Wronski, A. R. Badzian, T. Badzian, R. Messier, "The Effect of Surface Treatment on the Electrical Properties of Metal Contacts to Boron-doped Homoepitaxial Diamond Film", Electron. Device Lett. 11 (2) (1990) 100.

[3]M. I. Landstrass, K. V. Ravi, Appl. Phys. Lett., "Resistivity of Chemical Vapor Deposited Diamond Films", 55 (10) 975-977 (1989).

[4]J. W. Vandersande, L. D. Zoltan, "Using High-Temperature Electrical Resistivity Measurements to Determine the Quality of Diamond Films", Diamond and Related Materials 4, 641-644 (1995).

[5]S. Albin and L. Watkins, Appl. Phys. Lett., "Electrical Properties of Hydrogenated Diamond", 56, 1454-1456 (1990).

[6]A. K. Kulkarni, K. Tey, and H. Rodrigo, "Electrical Characterization of CVD Diamond Thin Films Grown on Silicon Substrates", Thin Solid Films, 270, 189-193 (1995).

[7]B. Feigl, R. Kuhnert, M. Ben-Chorin and F. Koch, Appl. Phys. Lett., "Evidence for Grain Boundary Hopping Transport in Polycrystalline Diamond Films, 65, 371-373 (1994).

[8]Y. Muto, T. Sugino, and J. Shirafuji and K. Kobashi, "Electrical Conduction in Undoped Diamond Films Prepared by Chemical Vapor Deposition", Appl. Phys. Lett., 59 (7) 843-845 (1991).

[9]C. F. Chen, S. H. Chen, K. M. Lin, "Electrical Properties of Diamond Films Grown at Low Temperatures", Thin Solid Films 270, 205-209 (1995).

[10] A. T. Collins, "The Electrical and Optical Properties of Thin Film Diamond", Ceramics International 22, 321-7 (1996).

[11]V. Shanov, W. Tabakoff, and R. N. Singh, "CVD Diamond Coatings for Erosion Protection at Elevated Temperatures", Journal of Materials Engineering and Performance, 11, 2, 2002.
[12]V. Shanov, R. N. Singh, and W. Tabakoff, invited presentation at the International Conference on Metallurgical Coatings and Thin Films, April 10-14, 2000, San Diego, California.
[13]R. Ramamurti, V. Shanov, R. N. Singh, M. Samiee, and P. Kosel, "Polycrystalline Diamond Thin Films for High Temperature Electronics and MEMS", The Sixth Applied Diamond Conference/Second Frontier Carbon Technology Joint Conference Proceedings NASA, edited by Y. Tzeng, K. Miyoshi, M. Yoshikawa, M. Murakawa, Y. Koga, K. Kobashi, and G. A. J. Amaratunga, CP-2001-210948, 64-67, (2001).
[14]D. Zhou, D M. Gruen, L. C. Qin, T. G. McCauley, and A. R. Krauss, "Control of Diamond Film Microstructure by Ar Additions to CH_4/H_2 Microwave Plasmas, Journal of Applied Physics, 84, 4-10 (1998).
[15]S. M. Sze, Physics of Semiconductor Devices, 2nd ed. (Wiley, Newyork, 1981).
[16]B. R. Stoner, J. A. Von Windheim, J. T. Glass, "Electrical Conductivity as a Function of Temperature of Diamond Films Grown by Downstream Microwave Plasma Chemical Vapor deposition", Mat. Res. Soc. Symp. Proc., 270, 413-418 (1992).
[17]K. Okumura, J. Mort, and M. A. Machonkin, "Density of States Distribution in Diamond Thin Films", Appl. Phys. Lett., 59, 455-457 (1990).
[18]N. F. Mott and E. A. Davies, Electronic Processes in Non-crystalline materials (Clarendon, Oxford, 1971).

DIELECTRIC PROPERTIES OF SUSPENSIONS CONTAINING $BaTiO_3$ PARTICLES

Abhishek A. Manohar and Fatih Dogan
Department of Material Science and Engineering,
University of Missouri Rolla,
222 McNutt Hall,187□Miner Circle, UMR,
Rolla, MO 654□9.

ABSTRACT

The size effect of dielectric materials, grain size in sintered ceramics and particle size in powders, plays an important role in influencing their dielectric properties. The grain size effect differs from the particle size effect mainly because of the boundary conditions within a ceramic. The grains in a ceramic are three dimensionally bound by surrounding grains while a particle in a ceramic is free and not subjected to such microstructural effects. The dielectric constant of barium titanate powder compacts as well as slurries were measured using powders of various particle sizes. The calculation of dielectric constant of particles using finite element models were investigated with a critical discussion on the reliability between measured and calculated data.

KEYWORDS: Dielectric constant, BT slurry, finite element model.

INTRODUCTION

Grain size pays an important role in determining the dielectric properties of a ceramic[1]. Particle size also plays an important role in influencing the properties of the dielectric. As the particle size decreases to nanometer scale local microstructure and composition changes relative to the bulk state are greatly increased and especially at the particle surfaces. Hence the dielectric properties of the nanoparticles are quite different from the bulk-sized particles[2]. The dielectric constant and dielectric loss of barium titanate are very sensitive to changes in particle or grain size[3]. They are also dependant on the grain boundaries in case of sintered structures along with the domains and agglomerates. The effect of grain size on the dielectric properties of barium titanate has been studied and reported in several publications[1,3]. The effect of particle size on dielectric properties of barium titanate and the estimation of dielectric constant of a powder particle is still obscure.

The grain size effect differs from the particle size effect mainly because of the boundary conditions within a ceramic. The grains in a ceramic are three dimensionally bound by surrounding grains while a particle in a ceramic is free and not subjected to such restrictions, i.e. grains are subjected to effects by neighboring grains while particles are not. Mechanical stresses originate at the grain boundaries as a result of cubic to tetragonal transformation on cooling. Hence there is lot of stress on the grains in a sintered BT structure. These stresses are relaxed by the formation of complex domain structure[4].The size effects of particles in cluster form in barium titanate have also been reported[5]. These effects were deemed to be intermediate between behavior of single particle and single grain in a powder compact. The sizes of the particles also influence the room temperature crystal structure of barium titanate. Begg et al.[6] showed that particles less than 19□nm are cubic, while particles greater than 27□nm are tetragonal. It has also been suggested that barium titanate single crystalline grain consists of a cubic surface layer and tetragonal core.[7-8] Thus the corresponding dielectric behavior of barium titanate will depend on the surrounding conditions (mechanical stresses or not), grain and particle sizes. It has been reported that tetragonal barium titanate powders which are hydrothermally synthesized have better sintering behavior, higher density and higher dielectric constants than cubic phase powders[9]. Wada et al.[1□-11] has reported the dielectric constant of powder particle using finite element modeling and slurry measurements. The dielectric properties of unsintered barium titanate powders[12] have been studied and reported. Various models have been reported for the prediction of dielectric constant of ceramic-composites[13-18]. Yamada et al.[18] in 1982 had proposed a model for predicting the dielectric constant of □-3 composites where the ceramic particles are dispersed in a continuous medium Similarly a self consistent model was proposed by Rao[15] was based on the effective medium theory and was particularly modeled for nanostructured polymer-ceramic composites. But there are very few publications which take into account the dielectric constant of a powder particle. The intent of this study was to determine the dielectric constant of the powder particles. Most of the studies deal with dielectric properties of sintered compacts, but in this study the dielectric properties in unsintered condition were investigated.

EXPERIMENTAL PROCEDURE

The intent of this study was to determine the dielectric constant of barium titanate [BT] slurries and from that investigate the dielectric properties of individual BT particles. For this work commercially available barium titanate nanopowder HPB 1□□□ having particle size of 5□nm and barium titanate X7R having particle size of 1.15μm were used. These powders were homogenous, uniformly sized and of very high purity. Figures 1a and 1b show the SEM micrographs of nanosized powder HPB 1□□□ in as received and sintered condition respectively. Figures 1c and 1d show the micrographs for micron sized powder X7R. The micrographs suggest clustering of particles in case of as received powders which could affect the dielectric properties[5]. The procedure for the preparation of BT slurry included ball milling which ensured the complete and homogenous dispersion of BT particles. Figures 2a and 2b show the X-Ray diffraction peaks for barium titanate powders in as received state. Surface water on the nanoparticles plays a very important role in determining the dielectric properties of barium titanate[19]. Hence to completely remove the surface water, moisture the barium titanate powders were vacuum dried at 12□°C for 5 hrs.

These ceramic powders were pressed into discs with 4□±□□1 mm in diameter and 2mm in thickness using a die press. Dielectric properties of the unsintered powder pellets were measured. The disc samples were sintered at different temperatures and the dielectric constant for sintered powder compacts was determined. The optimum sintering temperature and sintering time which gave the highest dielectric constant for sintered discs was determined and the rest of the discs were sintered at the same temperature of 128□°C for 3 hours. The sintered compacts were then crushed to get the sintered powder. The average particle size of the sintered powder was between 15-2□μm. From the sintered powder, pellets were pressed and their dielectric constants were calculated. Next step involved the preparation of BT slurry. For the preparation of slurry the organic solvent chosen was propylene carbonate. This organic solvent was chosen due to its harmless nature towards $BaTiO_3$ and also a high dielectric constant of 66.5□ at room temperature of 2□°C. The barium titanate powders both sintered and as received were mixed in the propylene carbonate in increasing volume concentration and these mixtures were ball milled for 24 hours to form uniform BT slurry. The capacitance of the slurry was measured with the help of special attachment to the impedance analyzer. The dielectric loss values were noted. The dielectric constant of BT slurry was studied in combination with the proposed finite element models. All the samples prepared according to the ASTM standards with the diameter of pellet at 4□±□1 mm and thickness between 2 to 3mm. Proper care was taken to avoid errors like fringe effects, stray capacitance, etc during capacitance measurements.

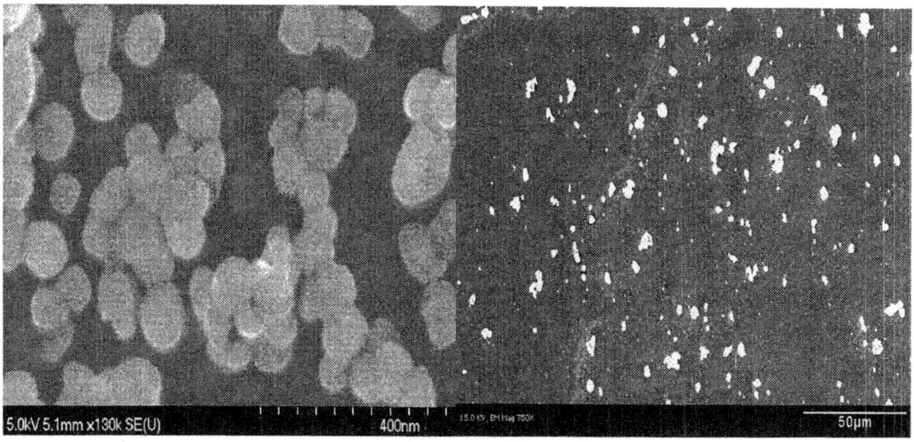

(a) As Received (b) Heat Treated

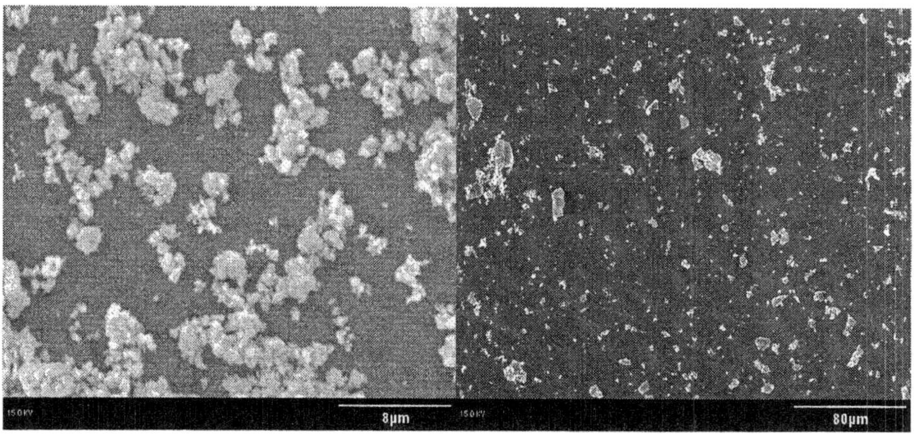

(c) As Received (d) Heat Treated

Figure 1: SEM images of (a) as received HPB 1□□,
(b) heat treated HPB 1□□,
(c) as received X7R,
(b) heat treated X7R.

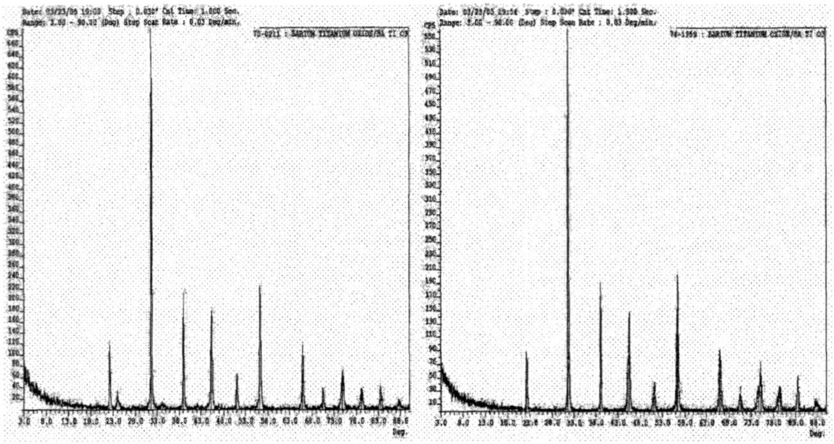

(a) BT X7R (b) HPB 1□□□

Figure 2: X-Ray diffraction peaks of $BaTiO_3$ for X7R and HPB 1□□□[nanosized powders].

RESULTS AND DISCUSSION

Powder Compacts

The capacitance values for all the powder compacts and slurries were measured using the impedance analyzer (HP 4192A LF impedance analyzer) at room temperature. From the capacitance data, the dielectric constant at room temperature was calculated using

$$e = C \times d / e_{0} A \tag{1}$$

where C is capacitance, D the thickness and e_{0} the free space permittivity and A the area of the pellet. The dielectric constant was determined at varying frequencies. Figures 3a and 3b show the variation of dielectric constant for X7R and nanosized powder HPB 1□□□ The dielectric constant for the as received powder compacts is slightly less than the dielectric constant of the sintered powder compacts, although the dielectric constant of the nanosized powder is greater than the micron sized X7R. The packing density for the X7R pellets was at 52% whereas for the nanosized powder it was at 6□%. The finer particle size allowed for higher packing density in case of nanosized powder.

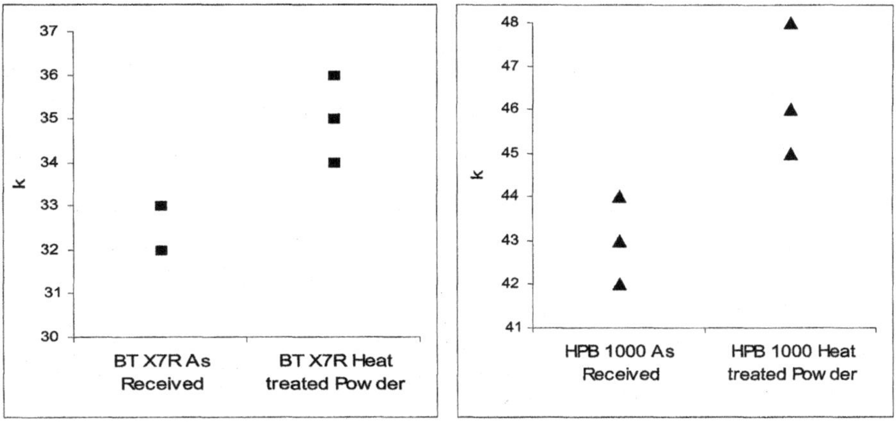

(a) (b)

Figure 3: Variation of dielectric constant for X7R and HPB 1000 powder compacts.

Slurry Data

The capacitance of slurry [BT in PC] was measured with a special attachment to the impedance analyzer HP 4192A LF. By using equation 1, the dielectric constant of BT slurry was calculated. The dielectric constant of pure propylene carbonate is known to be at 66.5 at room temperature. To validate our measurement technique and results the dielectric constant of propylene carbonate was measured using the special tool attached to the impedance analyzer. The measured dielectric constant of the propylene carbonate was within ± □□1 of the known value of 66.5. Figure 4 shows the variation of dielectric constant of slurry with increasing frequency. With increasing BT content, the dielectric constant of the slurry increases linearly as shown in Figure 5. The dielectric loss was also recorded and the Figure 6 shows the variation of dielectric loss with increasing frequency and increasing BT content for the slurry. Figure 7a shows the variation of dielectric constant of slurries for nanosized powder HPB 1000 (sintered and unsintered) while Figure 7b shows the similar data for micron sized X7R. The dielectric constant of the slurry containing sintered BT powder is higher than that of the slurry with unsintered [as received] powder in both the cases [HPB and X7R]. This increase in the slurry dielectric constant can be due to the increase in particle size or due to the sintering process which lead to changes in the microstructure of the barium titanate. The exact mechanism is not understood and needs to be investigated. The dielectric constant of powder compacts (sintered and unsintered powders) for HPB 1000 is greater than that for BT X7R while the slurry measurements indicate otherwise. This effect can be attributed to the higher packing density of HPB 1000 due to smaller particle size as compared to the larger particle sized X7R. Yamada[18] model was used to calculate the dielectric constant of the powder particle from the slurry data. Similarly a self consistent model proposed by Rao[15] was used to calculate the dielectric constant of BT particle. The self consistent model was limited to the nanoparticulate composites as reported by the authors.

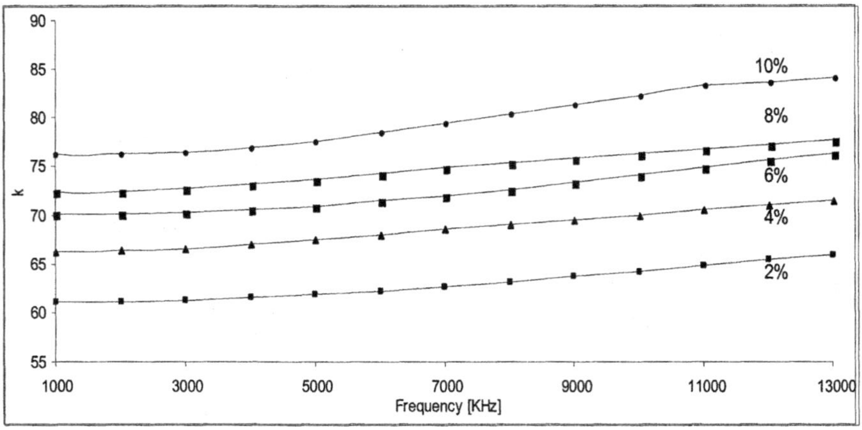

Figure 4: Variation of dielectric constant of slurries with increasing frequency for increasing volume % of nanosized BT.

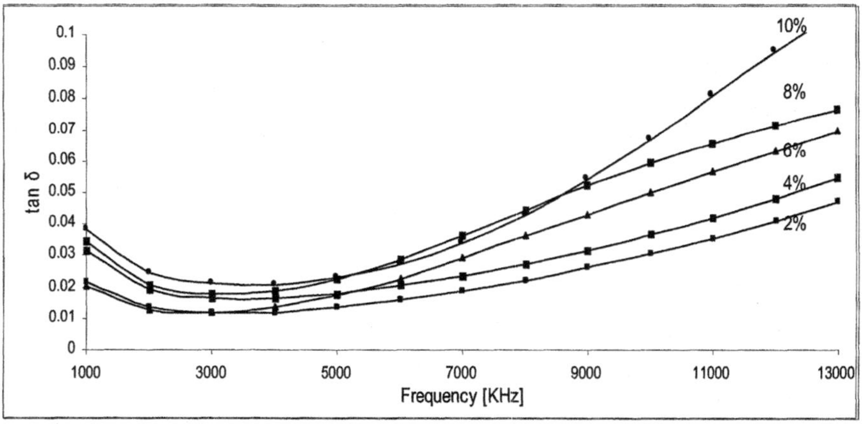

Figure 5: Dielectric loss variation with frequency for nanosized BT slurries with increasing volume % of nanosized BT.

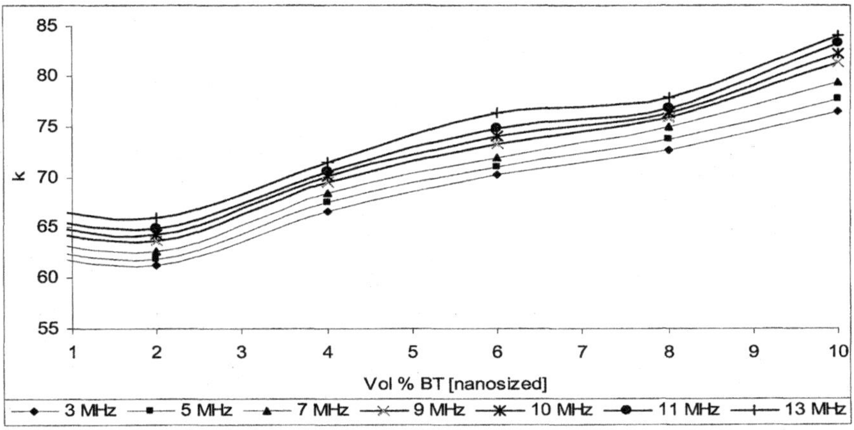

Figure 6: Effect of increasing volume % of nanosized BT on slurry dielectric constant for different frequencies.

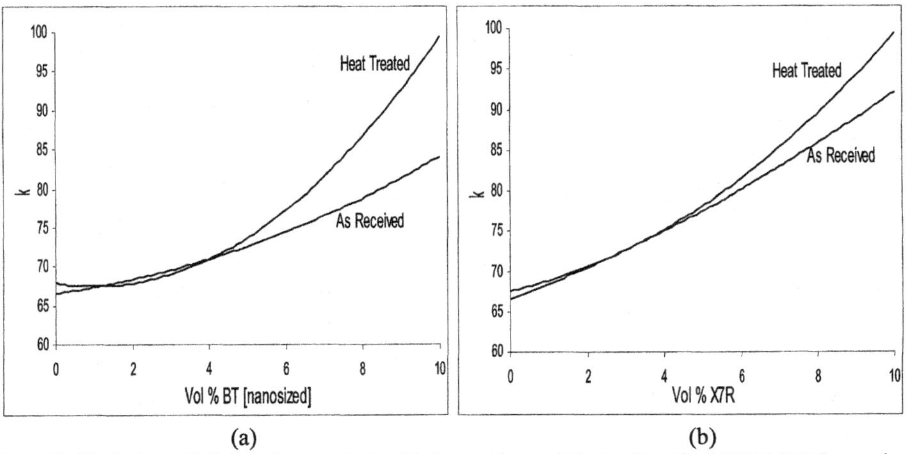

Figure 7: Variation of dielectric constant with increasing solids loading for HPB 1□□□[nanosized powder] and X7R.

These calculations revealed the sensitive nature of finite element models proposed. A small (fractional) change in the dielectric constant value of slurry led to a large change in the calculated dielectric constant of powder particle. Not only these models but several other models were investigated for the dielectric constant calculations. Similar results were encountered in all the cases. Figure 8a and 8b show the sensitive nature of models investigated. Calculations of

Figure 8 are based on equation used by Rao et.al[15]. The equation is also mentioned in the Figure 8a. Calculations were made for 2% BT slurry.

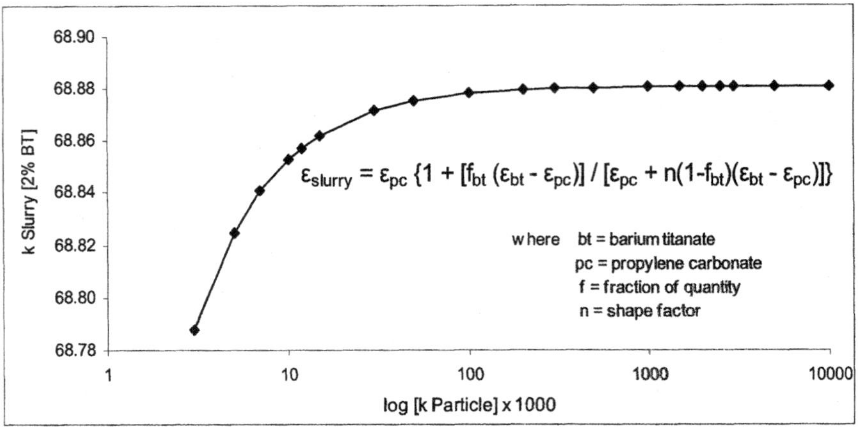

(a)

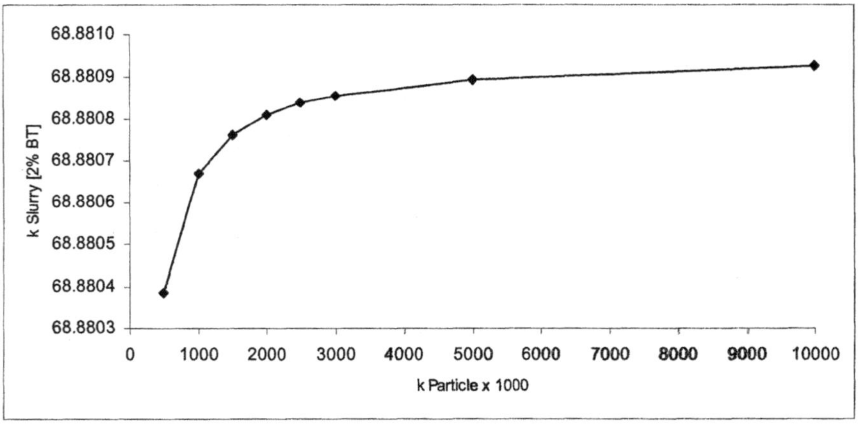

(b)

Figure 8: (a) Slurry dielectric constant variation as compared to powder particle (log scale). (b) Slurry dielectric constant variation as compared to powder particle for particle dielectric constant greater than 5□□□□(linear scale).

From the Figure 8, for every fractional change [1□^{2} to 1□^{4}] in the dielectric constant of slurry the dielectric constant of particle changes by several orders of magnitude. Hence due to the high dependence of the calculated value of dielectric constant of BT particle on the model itself and due to the model's sensitive nature, the dielectric constant of the powder particle could not be confidently reported at the time. The slurry data indicates that the dielectric constant of the sintered particles is greater than the dielectric constant of the unsintered particles. This effect can be due to the particle size as the particle sizes of sintered powders are larger than then unsintered powders. Or this effect can be due to the sintering process which lead to a change in the microstructure of the particle and enhanced the dielectric properties.

CONCLUSIONS

The dielectric properties of BT slurry as a function of frequency and BT powder content was measured and reported. The sintered powder slurries show a higher dielectric constant than the unsintered powder slurries whereas the powder compacts prepared from sintered and unsintered powders show similar dielectric constant values. This effect needs to be investigated and studied in detail. The finite models indicate a very high dielectric constant of BT powder particles. But due to the very sensitive nature of the models, the dielectric constant of powder particles could not be accurately calculated. Future theoretical and experimental work will include modeling for the slurries and microstructural examination of the powders in detail.

REFERENCES

[1] A.J. Bell, A.J. Moulson and L.E. Cross, "The effect of grain size on permittivity of $BaTiO_3$", Ferroelectrics, **54**, 147-15□(1984).

[2]S.Berger, "Dielectric Properties of Nanoparticles", Dekker Encyclopedia of nanoscience and nanotechnology, **2**, 371-387 (2□□4).

[3]Xiao-Hui Wang,Ren-Zheng Chen,Zhi-Lun Gui,Long-Yu Li, "The grain size effect on dielectric properties of BaTiO3 based ceramics", Materials Science and Engineering, **B99**, 199-2□2 (2□□3).

[4]G. Arlt, D. Hennings, and G. DeWith, "Dielectric properties of fine grained barium titanate ceramics," J.Appl. Phys., **58**, No. 4, 1619-1625(1985)

[5]X .Li, W.H. Shih, "Size effects in barium titanate particles and clusters", J.Am.Ceram.Soc, **80(11)**, 2844-52 (1997)

[6]B. D. Begg, E. R. Vance and J. Nowotny, "Effect of particle size on room temperature crystal structure of barium titanate." J.Am.Ceram.Soc, **77(12)**, 3186-92 (1994)

[7]T. Takeuchi, K. Ado, T. Asai, H. Kageyama, T. Saito, C. Masquelier, O. Nakamwa: "Thickness of cubic surface phase on barium titanate single crystalline grains". J.Am.Ceram.Soc. **77(6)**, 1665-68 (1994)

[8]J.C. Niepce, Dijon, Fr., "Some aspects of the influence of particle size on properties and behavior of dielectric materials: Example of barium titanate" NATO ASI Series, Series E: Applied Sciences, **173**, 521-33 (1989).

[9]W. Azu, S.A. Akbar, R. Asiaie, P.K. Dutta, "Sintering and Dielectric properties of hydrothermally synthesized cubic and tetragonal barium titanate powders." J.Appl.Phys, **36**, 214-21 (1997).

[1□]Wada, Satoshi; Yasuno, Hiroaki; Hoshina, Takuya; Nam, Song-Min; Kakemoto, Hirofumi; Tsurumi, Takaaki, "Dielectric properties of nm-sized barium titanate crystallites with various particle sizes," Ceramic Transactions, **150** (Ceramic Materials and Multilayer Electronic Devices), 2□5-219 (2□□4).

[11]Wada, Satoshi; Yasuno, Hiroaki; Hoshina, Takuya; Nam, Song-Min; Kakemoto, Hirofumi; Tsurumi, Takaaki, "Preparation of nm-sized barium titanate fine particles and their powder dielectric properties," Japanese Journal of Applied Physics, Part 1: Regular Papers, Short Notes & Review Papers, **42(9B)** , 6188-6195(2□□3).

[12]Goswami, Amiya K., "Dielectric properties of unsintered barium titanate." Journal of Applied Physics, **40(2)**, 619-24 (1969).

[13]K. Wakino, T. Okada, N. Yoshida, K. Tomono, "A new equation for predicting the dielectric constant of a mixture", J.Am.Ceram.Soc, **6(0),** 2588-94 (1993).

[14]Ceramic materials for electronics-processing, properties and applications- Edited by Relva C. Buchanan, MARCEL DEKKER, INC. 1986.

[15]Rao, Yang; Wong, C. P.; Qu, Jianmin; Marinis, Tom, "Self-consistent Model for Dielectric Constant Prediction of Polymer-ceramic Composite,", Proceedings - International Symposium on Advanced Packaging Materials: Processes, Properties and Interfaces, Braselton, GA, United States, Mar. 6-8, 44-49 (2□□□).

[16]Huebner, W.; Jang, F. C.; Anderson, H. U.., "Dielectric and electrical properties of BaTiO3 composites," Materials Science Research, **20** (Tailoring Multiphase Compos. Ceram.), 433-43 (1986)

[17]Bottcher, C. J. F., "The dielectric constant of crystalline powders," Recueil des Travaux Chimiques des Pays-Bas et de la Belgique, **64**, 47-51(1945).

[18]Yamada, T., T. Ueda and T Kitayama, "Piezoelectricity of high-content lead zirconate titanate/Polymer composite", J. Phys. Phys., 53(4), 4328-4332 (1982).

[19]Ma, Dongling; Siegel, Richard W.; Hong, Jung-Il; Schadler, Linda S.; Martensson, Eva; Oenneby, Carina., "Influence of nanoparticle surfaces on the electrical breakdown strength of nanoparticle-filled low density polyethylene", Journal of Materials Research, **19(3)**, 857-863 (2□□4).

ENHANCEMENT OF CYSTAL GROWTH IN MELT TEXTURING Ca-DOPED Y-Ba-Cu-O SUPERCONDUCTORS

Oratai Jongprateep and Fatih Dogan
Department of Materials Science and Engineering
University of Missouri-Rolla
1870 Miner Circle
Rolla, MO 65409

ABSTRACT

Doping of Ca in $YBa_2Cu_3O_{7-x}$ (Y123) has been reported to enhance the critical current density in Y123. However, fabrication of large Ca doped Y123 single crystals remains challenging. The cooling temperature and the growth rate of Ca doped Y123 samples as a function of Ca concentrations is studied. Isothermal cooling at varied temperatures (1003, 1005, 1006 and 1009°C) is applied to Ca doped Y123 samples in order to determine the optimal growth condition of the samples. It is found that the optimal growth rates occur at the undercooling ranges from 10 to 30°C in samples with Ca concentration ranges from 7.5 to 25 mol%, respectively.

INTRODUCTION

With its unique electrical and magnetic properties, high temperature superconductor $YBa_2Cu_3O_{7-\delta}$ (Y123) has a great potential for practical applications. High critical current densities J_c in the Y123 superconductor, typically in range of 10^4-10^6 A/cm^2, are required for these applications. However, the attainment of high J_c has been hindered by the interface charging and the bending of the electronic band structure at the grain boundaries [1-2]. It is reported that the enhancement of the grain boundary critical current density in Y123 can be achieved through the doping of Ca [3-4]. Substitution of the Y sites by Ca is found to increase hole concentrations in the CuO_2 plane. This effect can result in an oxygen over-doped regime at the grain boundaries. Therefore, the Ca doped Y123 system is an interesting system for the study of the superconducting properties in order to obtain high J_c values.

Effect of Ca doping on the superconducting properties of Y123 has been reported previously [5-9]. However, the relationship among the Ca concentration, the processing parameters and the crystal growth of melt textured Y123 doped with Ca has not been well established. According to the solute diffusion-limited model, formation of Y123 occurs following the peritectic reaction [10]:

$$Y_2BaCuO_5\ (s) + 3BaCuO\ (l) + 2\ CuO\ (l) \Rightarrow 2\ YBa_2Cu_3O_{6.5} \qquad (1)$$

To allow the reaction to proceed continuously, Y211 phase must readily dissolve in the Y123 liquid phase. Since undercooling is one of the most crucial factors that control the dissolution rate of the Y211 particles, it is important to optimize the undercooling in order to enhance the growth rate of the Y123 melt textured crystal. This study addresses the effect cooling temperature and Ca concentration on Y123 crystal growth rate.

EXPERIMENTAL PROCEDURE

Ca doped Y123 powders with the compositions of $Y_{0.925}Ca_{0.075}Ba_2Cu_3O_{7-x}$(A), $Y_{0.85}Ca_{0.15}Ba_2Cu_3O_{7-x}$(B), $Y_{0.8}Ca_{0.2}Ba_2Cu_3O_{7-x}$(C) and $Y_{0.75}Ca_{0.25}Ba_2Cu_3O_{7-x}$(D) were prepared by the solid state reaction technique from commercially available Y_2O_3, CuO, $BaCO_3$ and $CaCO_3$ powders (Alfa Aesar, Wardhills, MA, USA). After three cycles of calcinations at 890, 890 and 920°C with intermediate grinding, the powders were uniaxially pressed at 100 MPa to obtain pellets with 25 mm diameter.

Pellets were placed on substrates made of barium zirconate and heated from room temperature to 980°C with the heating rate of 300 °C/hr. The samples were subsequently heated and held at 1040°C for 2 hours. Isothermal cooling at the temperature of 1003°C, 1005°C, 1006°C, and 1009°C was applied for 15 hours to each set of samples; A, B, C and D, respectively.

Differential thermal analysis (DTA: Netzsch STA 409C/CD) was conducted to estimate the decomposition temperature of each powder composition. The measurement was conducted under the range of temperature from room temperature to 1100°C. Growth rates of the samples were calculated from the holding time at the growing temperature and the crystal size of samples, which was determined by measuring a distance from corner to corner of the crystal growth region.

RESULTS

The addition of Ca in Y123 superconductors demonstrates its effect on the increase of the decomposition temperatures. Figure 1 shows the DTA curves measured from 950 to 1100°C under a flowing air atmosphere of samples with Ca addition in the amount ranged from 7.5 to 25 mol%.

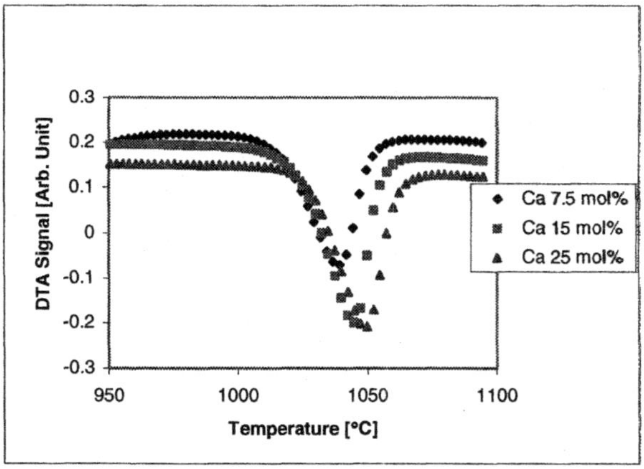

Figure 1. DTA profile for Ca doped powder with the compositions of A, B and D

As shown in Figure 2, the decomposition temperatures of the samples are 1016.5, 1023.4, 1025.5 and 1027.8°C in the samples with 7.5, 15, 20 and 25 mol% Ca, respectively. The results reveal that the decomposition temperature increase as high as 11°C, as the Ca doping concentrations increase from 7.5 to 25 mol%. This is in agreement with the earlier studies that the peritectic decomposition temperature of Ca doped Y123 is higher than that in the undoped material [5], [8].

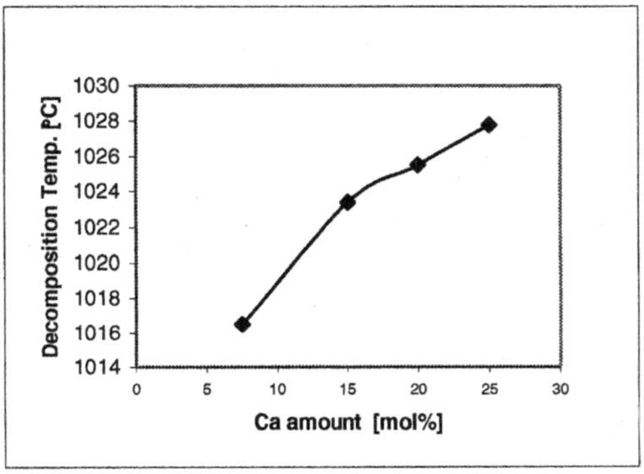

Figure 2. Decomposition temperature of Ca doped YBCO as a function of doping amount

A systematic study of the growth rates as a function of Ca concentration is shown in Figure 3. The isothermal temperatures of 1003 and 1009°C did not seem to promote any crystal growth. On the other hand, crystal growth was observed in samples held isothermally at 1005 and 1006°C. The Y123 single grains with a square planar growth morphology were obtained in the samples with smaller amounts of Ca (7.5 and 15 mol%). The growth rates of these samples were determined from the crystal sizes and the holding time at growth temperatures. It was found that growth rates of the samples were in the range of 0.2-0.25 mm/hr on the a-b plane.

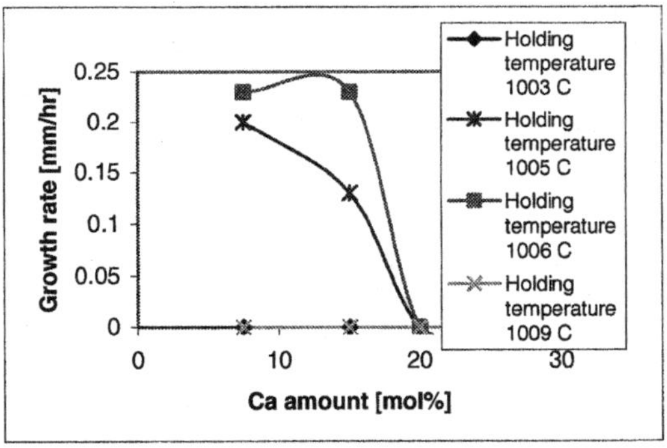

Figure 3.Growth rate as a function of doping amount of samples held at 1003, 1005, 1006 and 1009°C respectively

The crystal growth rate largely depends on doping amount of the samples and undercooling. For samples with lower amounts of Ca, lower undercooling is required to obtain larger Y123 crystals. Figure 4 shows the growth rate as a function of undercooling of samples with different amounts of Ca concentrations. The results indicate that the optimal growth rate occur at ΔT = 16.5°C, 23.4°C, 26.5°C and 28.8 °C for sample with Ca in the amount of 7.5, 15, 20 and 25 mol%, respectively.

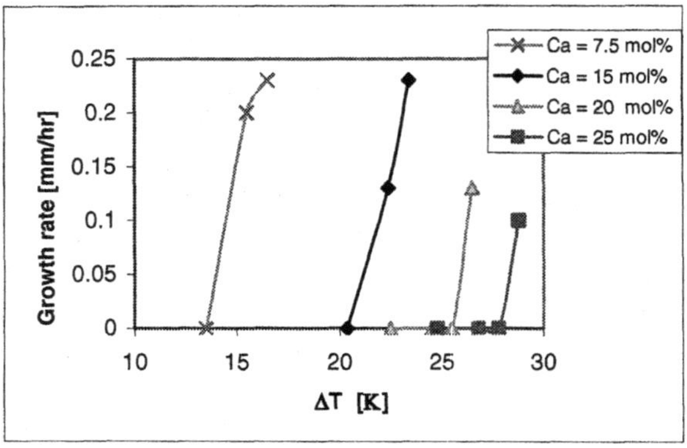

Figure 4.Growth rate as a function of undercooling temperature with different amounts of Ca concentration

DISCUSSION

Crystal growth experiments of the Y123 samples with various amount of Ca doping were carried out under the condition of isothermal cooling. The experimental results indicate a strong influence of undercooling temperature and sample compositions on the crystal growth. The decomposition temperatures of samples increase significantly as the Ca concentration increases. The modification of decomposition temperatures affects crystal growth temperature in the melt texturing process. Since undercooling is defined as the difference between the growth temperature (isothermal holding temperature) and the peritectic decomposition temperature, the growth temperatures need to be adjusted accordingly, in order to promote the optimal growth rate of Ca doped Y123 crystals.

An increase of Yttrium ion concentration in the Ba-Cu rich liquid phase is required for the formation of Y123, which occurs following the peritectic reaction (shown in Equation 1). To allow the reaction to proceed continuously, Y211 phase must readily dissolve in the liquid phase. The solute diffusion-limited model for peritectic solidification is described here to interpret the experimental results. The model assumes limited diffusion of Y211 in the liquid phase during the Y123 formation [11]. Therefore, the solidification process in the melt-textured growth of Y123 largely depends on the dissolution rate of the Y211 particles.

Undercooling is a crucial driving force for the Y211 dissolution during solidification of Y123. The diffusion flux from a Y211 particle can be described in the following equation [10]:

$$J = D/z \, [\, 1/m^{211}(2\Gamma/r + Gz) + (1/m^{123} - 1/m^{211})(\Delta T) \,] \qquad (2)$$

where J is the diffusional flux, D is the diffusivity, z is the distance from the solidification interface, m^{211} is the liquidus slope of Y211 phase, m^{123} is the liquidus slope of Y123 phase, G is the temperature gradient, Γ is the Gibbs-Thomson coefficient, r is the radius of Y211 particles, and ΔT is the constitutional undercooling.

The constitutional undercooling, corresponds to the temperature difference between the solidification interface and the actual temperature at the distance z, strongly relates to the undercooling (defined previously). Equation 2 then indicates that the flux necessary for the Y123 crystal growth is dependent on the undercooling. In this set of experiments, growth rates are found to increase as the undercooling increases. The degree of undercooling that promotes the optimal crystal growth varies with sample compositions. Higher undercooling is desirable in the growth process of the Y123 with larger Ca concentrations. For instance, the undercooling of 16.5°C is suitable for the Y123 sample with 7.5 mol% Ca, while the undercooling of 28.8 °C is needed to promote crystal growth in the sample with 25 mol% Ca.

Variation of the suitable undercooling as a result of the Ca doping has not been well explained. However, it is proposed that additions of Ca result in the modification of parameters given in Equation 2. Higher Ca concentration in Y123 may result in the modification of diffusivity, liquidus slope of Y211 and Y123 phase and Gibbs-Thomson coefficient. Further systematic studies are required to determine the effect of Ca on thermodynamics parameters of Y123.

SUMMARY

The relationship between the Ca amount, undercooling, and growth rate of Ca doped Y123 was investigated. It was observed that the doping amounts and the solidification temperature affect the crystal growth significantly. Larger undercooling is desirable in the growth process of the Y123 doped with larger Ca concentration. The optimal growth rates occur at the undercooling ranges from 10 to 30°C in samples with Ca concentration ranges from 7.5 to 25 mol%, respectively.

ACKNOWLEDGEMENTS

This work was supported by Boeing Phantom Works and the Materials Research Center at the University of Missouri-Rolla.

REFERENCES

[1]J. Mannhart, and H. Hilgenkamp, "Wavefunction symmetry and its influence on superconducting devices," *Supercond. Sci. Technol.*, **10**, 880-883 (1997).

[2]A. Gurevich, and E. A. Pashitskii, "Current transport through low-angle grain boundaries in high-temperature superconductors," *Phys. Rev. B*, **57**, 13878-13893 (1998).

[3]G. Hammerl, A. Schmehl, R. Schulz, B. Goetz, H Blelefeldt, C. Schnelder, H. Hilgenkamp, and J. Mannhart, "Enhanced supercurrent density in $YBa_2Cu_3O_{7-\delta}$ polycrystalline at 77k from calcium doping of grain boundaries," *Nature*, **407,** 162-164 (2000).

[4]A. Berenov, C. Farvacque, X. Qi, J. MacManus-Driscoll, D. MacPhail, and S. Foltyn, "Ca doping of YBCO grain boundaries," *Physica C*, **372,** 1059-1062 (2002).

[5]N. Hari Babu, M. Kambara, J. McCrone, J. Cooper, J. Tallon and D. Cardwell, "Fabrication of Ca-doped large grain Y-Ba-Cu-O superconductors," *IEEE Trans. Appl. Supercond.*, **11**, 3521-3524 (2001).

[6]C. Lin, B. Liang, and H. Chen, "Top seeded solution growth of Ca-doped YBCO single crystals," *J. Crys. Growth*, **237,** 778-782 (2002).

[7]G. Baldha, I. Shaikh, D. Kuberkar, J. Bhalodia, U. Joshi, and R. Kulkarni, "Enhanced flux pinning and critical currents by Hf and Hf-Ca substitution in $YBa_2Cu_3O_{7-\delta}$," *Cryogenics*, **35,** 61-65 (1995).

[8]F. Delorme, C. Harnois, I. Monot, and S. Marinel, "Ca doping of TSMTG-$YBa_2Cu_3O_{7-\delta}$ / Y_2BaCuO_5 composites," *Physica C*, **382,** 415-421 (2002).

[9]V. Awana, S. Malik, and W. Yelon, "Structural aspects and superconductivity in oxygen-deficient $Y_{1-x}Ca_xBa_2Cu_3O_{7-y}$ ($y \approx 0.3$) system. A neutron-diffraction study," *Physica C*, **262**, 272-278 (1996).

[10]Y. Shiohara and A. Endo, "Crystal growth of bulk high-T_c superconducting oxide materials," *Mater. Sci. Eng.*, **R19**, 1-85 (1997).

[11]M. Cima, M. Flemings, A. Figueredo, M. Nakade, H. Ishii, H. Broody, and J. Haggerty, "Semisolid solidification of high temperature superconducting oxides," *J. Appl. Phys.*, **72**, 179-190 (1992).

MICRO-RAMAN SPECTROSCOPY OF A VICKERS INDENT ON SOFT PZT

Jacob L. Jones and Mark Hoffman
Materials Science and Engineering
University of New South Wales
Sydney, NSW, 2052, Australia

ABSTRACT

Raman scattering has proven valuable in eliciting phase transitions in lead zirconate titanate ceramics. However, its use as a quantitative tool for spatially characterizing local strains and texture has yet to be fully optimized. This paper demonstrates that domain switching and microstrain surrounding a Vickers indent are manifested as changes in mode-specific intensities and Raman shifts. The intensity of the 265 cm^{-1} and 600 cm^{-1} modes increase inside the indent as a result of an increase in orientations of the corresponding bonds that are parallel to the indented surface.

INTRODUCTION

Raman scattering is a non-destructive tool for characterizing texture, strain, and composition in polycrystalline ceramics. The intensity of the scattered Raman spectra describes the change in polarizability of crystallographic bonds that correspond to a certain characteristic frequency or mode (cm^{-1}). The Raman modes expressed in polycrystalline ceramics are an average of those directionally-dependent modes exhibited by their single crystal counterparts. As such, the continuum between single crystal and polycrystalline Raman modes can indicate crystallographic texture,[1,2] analogous to that observed by relative intensity shifts in x-ray diffraction spectra.[3] Similarly, the shifting of Raman modes in the frequency spectrum (Δcm^{-1}) suggests microstrains, a more widely-employed use of the technique particularly in structural ceramics.[4]

The objective of this paper is to demonstrate the feasibility of using micro-Raman spectroscopy as a tool for spatially characterizing both domain switching and strain in ferroelectric ceramics. As a proof geometry, a Vickers indentation will allow significant domain switching and strains to be achieved and therefore the largest differences in intensities and peak shifts to be measured.[5]

The geometry of Raman scattering is illustrated in Fig. 1a. E_X is the polarized electric-field vector direction (X-direction) of the incident beam. The incident beam interacts with bonds oriented *in-plane* (X-Y plane). Those bonds oriented at an angle α to X scatter with intensity proportional to $\sin^2\alpha$.[6] The electric-field vectors of the scattered light are in all in-plane directions. This geometry is ideal for describing in-plane versus out-of-plane bonds and provides a greater scattered intensity than attainable by using a reflected beam polarizer.

The indentation geometry is illustrated in Fig. 1b. A random distribution of domain orientations are expected outside the indent, as in a typical unpoled ceramic. Inside the indent, in-plane domain orientations are preferred due to ferroelastic, stress-induced domain switching.

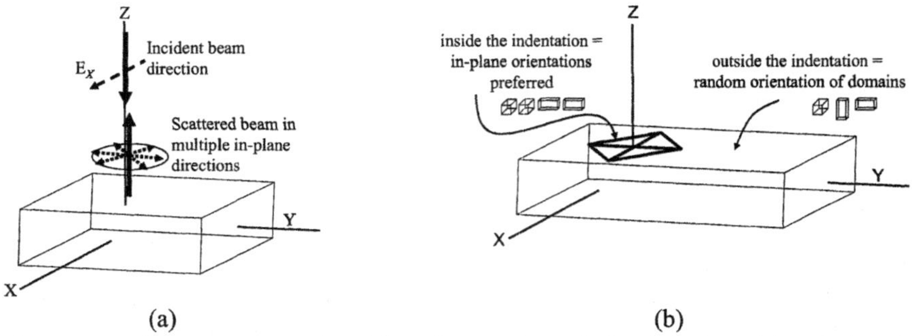

Fig. 1. Geometry of Raman scattering (a) and geometry of the indent and ideal domain orientations (b).

EXPERIMENTAL

A soft, commercial PZT (K350, Piezo Technologies, Indianapolis, Indiana) was used for this study. Indents were made on a polished and annealed surface using a 500 g Vickers indent and Raman spectra were obtained within 4 hours after indentation. A Renishaw 2000 Raman microscope, equipped with a polarized red incident laser (λ = 633 nm) was used in confocal mode with a 2 micron spot size. The measured spectra were unpolarized to obtain greater intensity (less noise) and because this geometry sufficiently measures the desirable domain orientations (Fig. 1). The measured spectra were fit with Gaussian profile peak shapes to obtain integrated areas and peak center positions.

Raman spectra were collected at various locations with respect to the indent and were also mapped in 2 μm step sizes along a 100 μm line originating at the center of the indent. These locations are identified in Fig. 2 using letters and a continuous line.

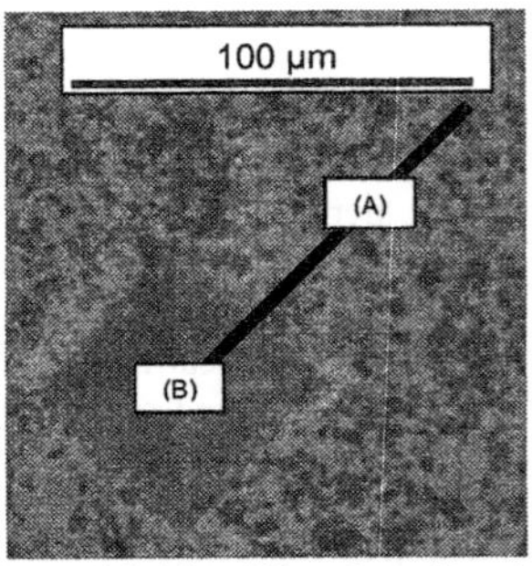

Fig. 2. Optical micrograph of indent with identifiers locating the positions of sample spectra and the line indicating the mapping direction.

RESULTS

The spectra obtained from two different locations with respect to the indent are shown in Fig. 3. Inside the indent, the intensity of the 265 cm^{-1} and 600 cm^{-1} modes increases relative to the random orientations outside the indent.

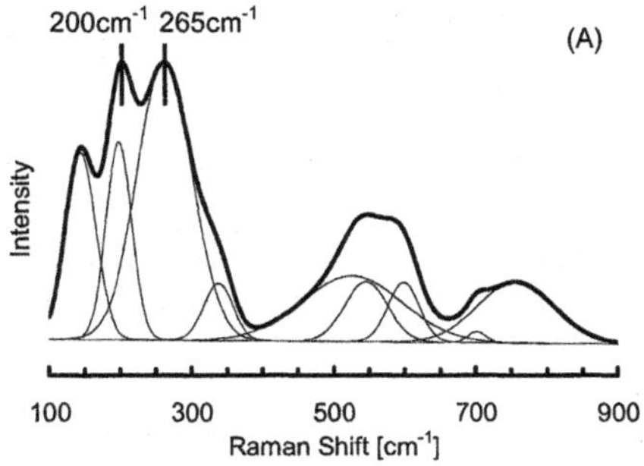

Fig. 3. Micro-Raman spectra at positions A (outside the indent) and B (inside the indent), previously identified in Fig. 2.

The change in intensity of the 265 cm^{-1} mode relative to the 200 cm^{-1} mode along a line from the center of the indent (point B in Fig. 2) to outside the indent (point A in Fig. 2) is shown in Fig. 4. The intensity of the 265 cm^{-1} mode is consistently larger within the indent. The shifts of the Raman modes are shown in Fig. 5, and are also different inside and outside of the indent.

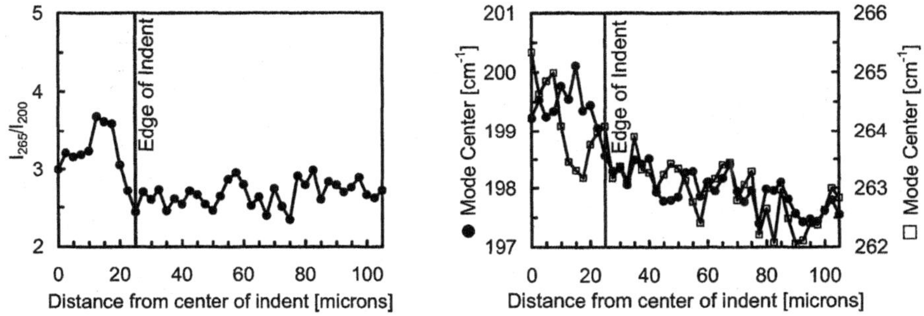

Fig. 4. Integrated intensity ratios and center positions of the 200 and 265 cm^{-1} modes as a function of distance from the center of the indent along line A-B in Fig. 2. Edge of the indent is also identified.

Fig. 5 shows results from the same mapping procedure after annealing the sample at 500°C for 2 hours. Some of the stresses in the indent expressed in the position of the 265 cm^{-1} peak are relieved by annealing. However, the stresses expressed in the position of the 200 cm^{-1} peak increase after annealing. Additionally, the intensity of the 200 cm^{-1} does not return to the random orientation intensity after annealing.

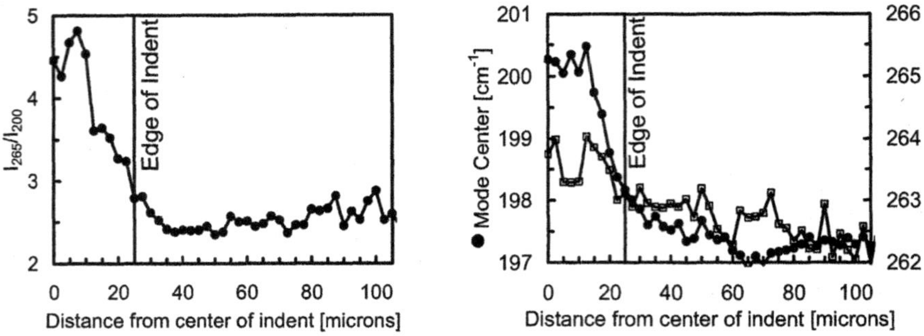

Fig. 5. Similar data as presented in Fig. 4, after the indent is annealed at 500°C.

DISCUSSION

Previously, it has been shown that the intensity of the 290 cm^{-1} mode is dependent upon the orientation of a tetragonal $PbTiO_3$ crystal whereas the 220 cm^{-1} is constant.[7] Given that these modes correspond to the same crystallographic bonds as the 265 and 200 cm^{-1} modes observed in ceramic PZT[8,9], these results are consistent. Within the indent, the quantity of domain orientations with the c-axis oriented in-plane is increasing and directly influences the observed intensities of the 265 cm^{-1} mode relative to the 200 cm^{-1} mode which is independent of the domain orientations. The appearance of the modes at a lower frequency in ceramic PZT compared to single crystal $PbTiO_3$ is not unexpected given the compositional and intergranular stress differences.[8-10]

After annealing, the domain orientations are not random because the plastic indent maintains a macroscopic boundary condition. The intergranular stresses expressed by selected modes (e.g., 265 cm^{-1}) are relieved during annealing while others (e.g., 200 cm^{-1}) are not relieved and even possibly increase.

CONCLUSIONS

Micro-Raman spectroscopy can measure domain switching and residual stresses on the micron size scale. Using this tool, it is observed that domain switching and residual stresses are present inside a Vickers indent on PZT. Because the plastic indent remains during annealing and provides a rigid boundary condition on cooling, domains are not completely randomized after annealing above the Curie temperature.

ACKNOWLEDGMENTS

This research was supported in part by the U.S. National Science Foundation under the International Research Fellowship Program, award number OISE-0402066. The support of Dr. Yu Wang, A/Prof. Alan Crosky, David Ruys, and Soodkhet Imlao at the University of New South Wales is gratefully acknowledged.

REFERENCES

[1]N. Dieckmann, R. Kürsten, M. Löhndorf, and A. Bock, "Epitaxial quality of c-axis and a-axis oriented $YBa_2Cu_2O_7$ films Characterization by Raman spectroscopy," *Physica C*, **C245**, 212-218 (1995).

[2]T. Puig, A. Puig-Molina, N. Mestres, H. V. Seijen, F. Alsina, J. C. González, X. Obradors, H. Graafsma, A. Usoskin, and H. C. Freyhardt, "Texture analysis of coated conductors by micro-Raman and synchrotron x-ray diffraction," *Mat. Res. Soc. Symp. Proc.*, **659**, II5.6/1-II5.6/6 (2001).

[3]J. L. Jones, E. B. Slamovich, and K. J. Bowman, "Domain texture distributions in tetragonal lead zirconate titanate by x-ray and neutron diffraction," *J. Appl. Phys.*, **97**, 034113-1-034113-6 (2005).

[4]G. Pezzotti, "Microprobe piezo-spectroscopy for the micromechanical analysis of fracture and deformation phenomena in polycrystalline ceramics," *Composites Science and Technology*, **59**, 821-831 (1999).

[5]T. Juliano, V. Domnich, and Y. Gogotsi, "Examining pressure-induced phase transformations in silicon by spherical indentation and Raman spectroscopy: a statistical study," *J. Mater. Res.*, **19**, 3099-3108 (2004).

[6]Z. Q. Lu, T. Quinn, and H. S. Reehal, "Polarization-dependent Raman spectra of thin crystalline silicon films," *J. Appl. Phys.*, **97**, 033512-1-033512-6 (2005).

[7]Z. Li, C. M. Foster, X.-H. Dai, X.-Z. Xu, S.-K. Chan, and D. J. Lam, "Piezoelectrically-induced switching of 90° domains in tetragonal $BaTiO_3$ and $PbTiO_3$ investigated by micro-Raman spectroscopy," *J. Appl. Phys.*, **71**, 4481-4486 (1992).

[8]G. Burns and B. A. Scott, "Raman spectra of polycrystalline solids; application to the $PbTi_{1-x}Zr_xO_3$ system," **25** [17], 1191-1194 (1970).

[9]S. J. Webb, N. Mcn. Alford, M. Poole, and M. J. Reece, "Raman characterisation of ceramic lead zirconate titanate," *Ferroelectrics 2000 UK*, 161-164 (2000).

[10]Q. Ma and D. R. Clarke, "Stress measurement in single-crystal and polycrystalline ceramics using their optical fluorescence," *J. Am. Ceram. Soc.*, **76**, 1433-40 (1993).

R-CURVE AND STRESS-STRAIN BEHAVIOR OF HARD AND SOFT PZT CERAMICS

Jacob L. Jones and Mark Hoffman
Materials Science and Engineering
University of New South Wales
Sydney, NSW, 2052, Australia

William F. Shelley, II.
Piezo Technologies
5460 West 84th Street
Indianapolis, IN 46268

ABSTRACT

The change in fracture toughness as a function of crack extension (R-curve) is measured and compared in both hard and soft lead zirconate titanate (PZT) ceramics using compact tension geometry. The soft PZT ceramic exhibited an initial fracture toughness of 0.70 MPa·m$^{1/2}$ and an increase in toughness of 0.66 MPa·m$^{1/2}$ while the hard PZT ceramic exhibited an initial fracture toughness of 0.78 MPa·m$^{1/2}$ and an increase in toughness of 0.10 MPa·m$^{1/2}$. The distinctive R-curve behavior of each composition is interpreted with respect to mechanical stress-strain results obtained using 4-point bending geometry. The intrinsic fracture toughness and extension toughening enhancement are correlated with bending strength and remanent strain, respectively.

INTRODUCTION

Lead zirconate titanate (PZT) ceramics exhibit increases in fracture toughness with crack extension, or R-curve, behavior. Such toughness increases are akin to phase transformation toughening of zirconia ceramics. In zirconia, the toughness increase results from a dilation strain in the frontal zone which slows crack propagation by crack-wake closure stresses.[1] In PZT, the accepted mechanism for the toughness increase is ferroelastic domain switching in the frontal zone creating crack-closure stresses in the crack wake during crack advancement.[2-10]

Much experimental work has thus far been reported in order to resolve this damage evolution mechanism in PZT. Lucato *et al.*[2] reported the effect of poling direction, Glazounov *et al.*[3] measured domain switching on the fractured surfaces using X-ray diffraction, Lucato[4] measured the influence of crack growth velocity, Karastamatis *et al.*[5] determined the influence of grain size, Oates *et al.*[6] determined the influence of electrical boundary conditions and humidity, and Kounga Njiwa *et al.*[7] have measured the geometry of the switching zone *in situ*. Most recently, Salz *et al.*[8] have studied this behavior under mechanical cyclic fatigue. Significant modeling efforts have also been undertaken to predict the fracture toughness enhancement as a function of material parameters.[9]

However, with the exception of some recent work on electrostrictive lead lanthanum zirconate titanate (PLZT)[10], the comparative R-curve behavior of more than one composition is rarely explored. No comparative study of the R-curve behavior of various PZT compositions has yet been presented. However, such an examination is critical to fully understanding the effects of stress-induced domain switching (ferroelasticity) on R-curve behavior and confirming the recently-developed constitutive models.[9]

The objective of this paper is to present the differences in R-curve behavior of hard and soft PZT ceramics and interpret these differences with respect to mechanical stress-strain behavior and the generally-accepted fracture mechanisms. As a secondary objective, this paper will demonstrate the necessity of various mechanical tests in interpreting fracture toughness results of ferroelastic ceramics.

EXPERIMENTAL

Soft and hard PZT compositions were obtained from Piezo Technologies (Indianapolis, IN), manufactured to the two desired geometries. The soft PZT ceramic (K350) is niobium-doped, the hard PZT ceramic (K270) is strontium-doped, and both compositions are slightly on the tetragonal side of the morphotropic phase boundary. Both ceramics have a grain size of ~2 μm.

Bend bars of both compositions were obtained from the supplier in dimensions 3×4×45 mm. Strain gauges (CEA-13-062UW-350, M-M, Raleigh, NC) were attached to the two 4×45 mm sample faces. The samples were mounted in a 4-point bending fixture with inner and outer spans of 20 and 40 mm. In this configuration, the strain gauges enable simultaneous measurement of strain on the tensile and compressive surfaces. Mechanical load was applied using an MTS810 servo-hydraulic mechanical test machine. The target surface stress, or bending stress, was calculated from the applied load (P), loading geometry (spans l_2 and l_1), and sample cross-section (width b, height h) using the equation

$$\sigma = \frac{\pm 3P(l_2 - l_1)}{2bh^2}. \tag{1}$$

Using displacement control, the samples were loaded and unloaded at rates corresponding to approximately 1.9 MPa/sec and 1.4 MPa/sec for the hard and soft PZT, respectively. The asymmetry in tension and compression of PZT ceramics may be analyzed to obtain the true applied surface stresses.[11] Here, this correction is unnecessary for subsequent interpretation of the results and is not undertaken.

Compact tension specimens of both compositions were obtained from the supplier in dimensions 50×48×3 mm with a notch 23 mm in length parallel to the 50 mm dimension. A chevron notch was cut at a 45° angle to the major surface of the sample at the end of the 23 mm long notch using a sharp razor blade with 6 μm diamond paste as a lubricant. A precrack was initiated on the major surface opposite the opening of the chevron notch (on the side of higher stress) using a 500 g Vickers indentation. The precrack was propagated using the controlled crack growth equipment described later for the fracture toughness measurements. After the precrack was propagated through the Chevron notch and was visibly present on both major surfaces, the samples were annealed for 2 hours at 600°C, above the ferroelastic-paraelastic transition temperature, to remove any effects of domain switching. A custom built apparatus was employed to measure fracture toughness. The compact tension specimens were loaded perpendicular to the crack face using a piezoelectric actuator. The load was recorded in-line with the actuator and sample using a 1 kN load cell. The crack length was measured initially and monitored continuously during crack growth using an optical microscope. The crack length, load, and sample geometry were all used to calculate the applied stress intensity factor using formulations for the compact tension geometry given elsewhere.[12,13] The applied stress intensity factor to maintain a crack velocity of approximately $1\cdot10^{-6}$ m/s in the subcritical crack growth regime was recorded as the fracture toughness as a function of crack extension, K_R.

RESULTS

The mechanical stress-strain behavior during loading to fracture is shown in Fig. 1. The ultimate bending strengths of the hard and soft PZT were 75 MPa and 50 MPa, respectively. There is a clear difference in the mechanical behavior of hard and soft PZT ceramics. The soft PZT is mechanically softer than the hard PZT and therefore exhibits a greater strain for a given stress. This observation is in agreement with similar reports using materials from different manufacturers[14] and is also analogous to that observed for the ferroelastic and electrostrictive compositions of PLZT ceramics.[10]

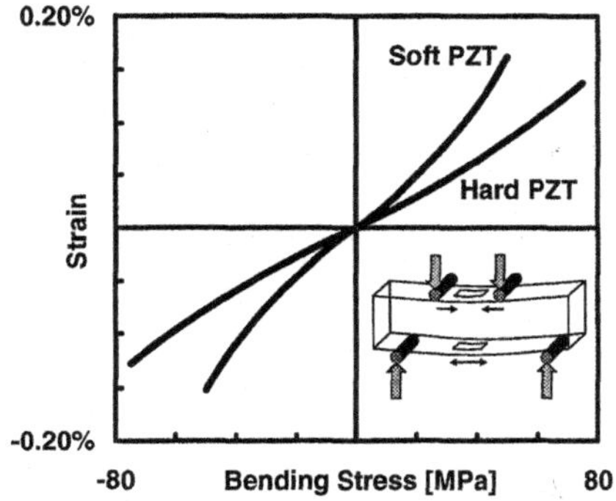

Fig. 1. Stress-strain behavior of hard and soft PZT ceramics tested in 4-point bend bar geometry to fracture. Tensile and compressive stress-strain behavior was recorded simultaneously.

When the ceramics are unloaded instead of fractured, the unloading path, hysteresis, and remanent strains can be recorded as shown in Fig. 2. The soft PZT ceramic exhibited a much larger hysteresis and remanent strain than the hard PZT ceramic. The measured remanent strains for the soft PZT ceramic in tension and compression were 0.024% and -0.022%, respectively. For the hard PZT ceramic, the remanent strains in tension and compression were ±0.004%, equivalent within the measurement error.

The two compositions exhibited significant differences in their R-curve behavior, as shown in Fig. 3. The soft PZT ceramic exhibited an initial toughness, $K_{r,o}$, of 0.70 MPa·m$^{1/2}$, an R-curve plateau, $K_{R,ss}$, of 1.36 MPa·m$^{1/2}$, and a rise distance, Δa_{ss}, of greater than 1 mm. The hard PZT ceramic exhibited values of $K_{r,o}$ = 0.78 MPa·m$^{1/2}$, $K_{r,ss}$ = 0.88 MPa·m$^{1/2}$, and Δa_{ss} = 40 μm. R-curve measurements were repeatable. Six different R-curve measurements of the soft PZT ceramic, annealing at 600°C between measurements, yielded an average $K_{r,o}$ of 0.73 MPa·m$^{1/2}$ and $K_{r,ss}$ of 1.38 MPa·m$^{1/2}$, both with standard deviations of 0.05 MPa·m$^{1/2}$. Both compositions exhibited transgranular fracture.

All of the results are summarized in Table I.

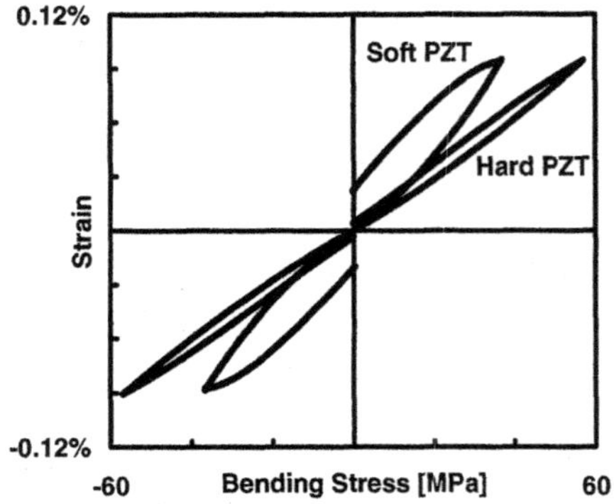

Fig. 2. Hysteretic stress-strain behavior of hard and soft PZT ceramics tested in 4-point bend bar geometry to a maximum of 75% of the bending strength determined in Fig. 1 and subsequently unloaded. Note different horizontal and vertical scales from Fig. 1.

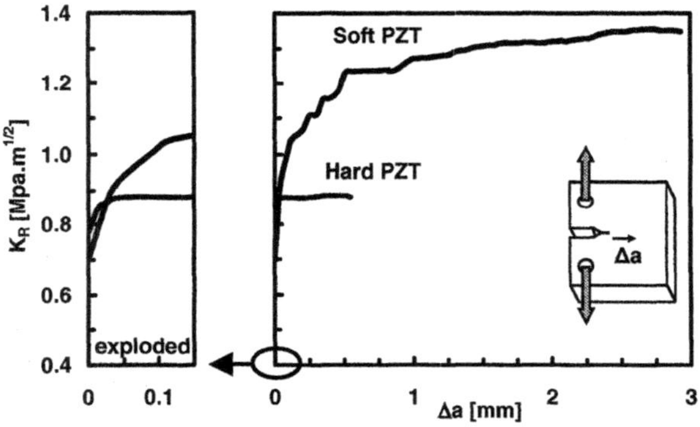

Fig. 3. R-curves of hard and soft PZT ceramics tested in compact tension geometry. The plot on left is an exploded view near the origin of the plot on the right.

Table I. Listing of the results obtained using 4-point bend bar and compact tension geometries.

	4-point bend bar			Compact tension [MPa·m$^{1/2}$]		
Material	Ultimate bending strength	Tensile remanent strain	Compressive remanent strain	Initial toughness, $K_{r,o}$	Steady-state toughness, $K_{R,ss}$	ΔK_R ($K_{R,ss}$ - $K_{r,o}$)
Hard PZT	75 MPa	0.004%	-0.004%	0.78	0.88	0.10
Soft PZT	50 MPa	0.024%	-0.022%	0.70	1.36	0.66

DISCUSSION

There is a significant difference in the R-curve behavior of hard and soft PZT ceramics (Fig. 3), the source of which can be elicited from the mechanical loading and bending strength results of Figs. 1 & 2.

Fracture strength is related to fracture toughness through[15]

$$K_{IC} = \sigma_f Y \sqrt{c}, \tag{2}$$

where c is the flaw size, Y is a geometrical parameter, σ_f is the fracture stress, and K_{IC} is the fracture toughness of the material. In materials which exhibit an R-curve, fracture strength is controlled by the rate of increase in toughness with crack extension relative to the increase in applied stress intensity factor. However, in structural ceramics, initial flaw sizes are usually small meaning that the effective toughness is essentially equivalent to the intrinsic toughness. In hard and soft PZT ceramics, a correlation exists between the initial fracture toughness ($K_{R,o}$) and bending strength: the soft PZT ceramic has a lower bending strength than hard PZT (50 MPa vs. 75 MPa) and a lower $K_{R,o}$ (0.70 MPa·m$^{1/2}$ vs. 0.78 MPa·m$^{1/2}$). Assuming equal flaw sizes, these correlations are reconciled using Eqn (2). More quantitatively, the measured values of each composition can be substituted into Eqn (2) yielding two equations. For constant Y, there are two unknowns which are the flaw sizes of each composition. Using the results obtained in this work as a first approximation, the flaw size of the soft PZT is expected to be approximately three times that of the hard PZT. An empirical measurement of the flaw size is unpractical as it is usually a single flaw at the fracture origin which defines the bending strength and *ex situ* observations are no doubt influenced by the fracture process. Further, such investigation is outside the scope of the current work – it has been shown that there is an order of magnitude agreement between different compositions.

Domain switching has long been considered the mechanism for crack extension toughening in ferroelastic materials.[2-10] This work demonstrates two PZT compositions with unique domain switching and toughening behavior. The remanent strains, a direct result of nonlinear domain switching, are approximately six times greater in the soft PZT composition than the hard PZT composition (Fig. 2). Greater remanent strains from domain switching provide greater crack extension toughening because domain switching shields the crack tip through crack wake closure stresses. Because hard PZT exhibits relatively little remanent domain switching compared to soft PZT, it is expected that hard PZT will have less crack wake closure stresses, less crack tip shielding, and therefore less extension toughening. The experimental results (Fig. 3) are in agreement with this hypothesis: the hard PZT exhibits a maximum increase in toughness, ΔK_R, of 0.10 MPa·m$^{1/2}$ while soft PZT exhibited ΔK_R = 0.66 MPa·m$^{1/2}$. These observations are consistent with recent constitutive model predictions[9], where it has been predicted that the toughness enhancement is inversely proportional to the material hardness and directly proportional to the ratio of irreversible strain to the characteristic elastic strain.

A closer examination of the initial fracture toughness values in light of the domain switching mechanisms should yield considerations that domain switching ahead of the crack tip will increase the observed initial toughness from the intrinsic material toughness. However, these effects are neglected in this analysis as a much more comprehensive quantitative approach would be necessitated for such an evaluation.

CONCLUSIONS

Bending strength and mechanical stress-strain behavior can be correlated with R-curve behavior in ferroelastic ceramics. To demonstrate, the R-curves of two PZT compositions with differing bending strength and mechanical behavior were measured. The bending strength and initial fracture toughness are related for constant flaw size. The increase in fracture toughness with crack extension (rising R-curve) results from domain switching in the crack wake. Soft PZT ceramics, with greater domain switching than hard PZT ceramics, exhibited a larger maximum increase in toughness, consistent with the widely-accepted active mechanisms.

ACKNOWLEDGMENTS

This research was supported by the U. S. National Science Foundation under the International Research Fellowship Program, award number OISE-0402066.

REFERENCES

[1]R. H. J. Hannink, P. M. Kelly, and B. C. Muddle, "Transformation Toughening in Zirconia-Containing Ceramics," *J. Am. Ceram. Soc.*, **83**, 461-87 (2000).

[2]S. L. dos Santos e Lucato, D. C. Lupascu, and J. Rödel, "Effect of Poling Direction on R-Curve Behavior in Lead Zirconate Titanate," *J. Am. Ceram. Soc.*, **83**, 424-26 (2000).

[3]A. E. Glazounov, H. Kungl, J.-T. Reszat, M. J. Hoffmann, A. Kolleck, G. A. Schneider, and T. Wroblewski, "Contribution from Ferroelastic Domain Switching Detected Using X-ray Diffraction to R-curves in Lead Zirconate Titanate," *J. Am. Ceram. Soc.*, **84**, 2921-29 (2001).

[4]S. L. dos Santos e Lucato, "Crack-Growth-Velocity-Dependent R-Curve Behavior in Lead Zirconate Titanate," *J. Am. Ceram. Soc.*, **86**, 1037-39 (2003).

[5]T. Karastamatis, D. C. Lupascu, S. L. dos Santos e Lucato, J. Rödel, and C. S. Lynch, "R-Curves of Lead Zirconate Titanate (PZT)," *J. Eur. Ceram. Soc.*, **23**, 1401-8 (2003).

[6]W. S. Oates, C. S. Lynch, D. C. Lupascu, A. B. Kounga Njiwa, E. Aulbach, and J. Rödel, "Subcritical Crack Growth in Lead Zirconate Titanate," *J. Am. Ceram. Soc.*, **87**, 1362-64 (2004).

[7]A. B. Kounga Njiwa, D. C. Lupascu, and J. Rödel, "Crack Tip Switching Zone in Ferroelectric Ferroelastic Materials," *Acta Mater.*, **52**, 4919-27 (2004).

[8]C. R. J. Salz, M. Hoffman, I. Westram, and J. Rödel, "Cyclic Fatigue Crack Growth in PZT Under Mechanical Loading," *J. Am. Ceram. Soc.*, in press (2005).

[9]C. M. Landis, "On the Fracture Toughness of Ferroelastic Materials," *J. Mech. Phys. Solids,* **51**, 1347-69 (2003).

[10]W. Chen, D. C. Lupascu, J. Rödel, and C. Lynch, "Short Crack R-Curves in Ferroelectric and Electrostrictive PLZT," *J. Am. Ceram. Soc.*, **84**, 593-97 (2001).

[11]T. Fett, D. Munz, and G. Thun, "Nonsymmetric Deformation Behavior of Lead Zirconate Titanate Determined in Bending Tests," *J. Am. Ceram. Soc.*, **81**, 269-72 (1998).

[12]ASTM standard E 399-90, *Annual Book of ASTM Standards.*

[13]ASTM standard E 561-98, *Annual Book of ASTM Standards.*

[14]J. Fan, W. A. Stoll, and C. S. Lynch, "Nonlinear Constitutive Behavior of Soft and Hard PZT: Experiments and Modeling," *Acta Mater.*, **47**, 4415-25 (1999).

[15]D. J. Green, *An Introduction to the Mechanical Properties of Ceramics*, Cambridge University Press, Cambridge, U.K., 1998.

Fuel Cells and Related Systems

FABRICATION OF SOFC ELECTRODES BY IMPREGNATION METHODS

Yingyi Huang, John M. Vohs, and Raymond J. Gorte
Department of Chemical & Biomolecular Engineering
University of Pennsylvania
311 Towne Building, 220 S. 33rd Street
Philadelphia, PA 19104

ABSTRACT

A novel procedure has been developed for fabricating composite electrodes for SOFC by impregnating a porous matrix of the electrolyte material with electronically conductive and catalytic materials. Because this method allows the electrolyte matrix to be calcined at high temperatures prior to adding the other components, thereby avoiding deleterious solid-state reactions, the method allows unprecedented compositional flexibility for both anodes and cathodes. This paper briefly reviews the present state of development for preparing SOFC through impregnation.

INTRODUCTION

The development of better electrodes remains an important research goal for solid oxide fuel cells (SOFC). For example, while the traditional Ni-based electrodes exhibit good performance in H_2, they are highly susceptible to carbon formation in the presence of hydrocarbons.[1] Indeed, Ni can catalyze the formation of carbon fibers even in the presence of syngas (mixtures of CO and H_2) under some conditions.[2] Cathodes are an even bigger problem since they often limit the performance of the SOFC. Because the cathode must have reasonably high electronic conductivity in air at high temperatures, the choice of electrode materials is limited to precious metals, which are impractical for cost reasons, and some electronically conducting oxides. The conducting oxides must not undergo reaction with the electrolyte and must have a reasonable CTE (coefficient of thermal expansion) match with the electrolyte.

In an SOFC electrode, the electrochemical reaction occurs at the three-phase boundary (TPB), that line where the electrolyte, the gas phase, and the electronic conductor meet. High performance electrodes are almost always porous composites of the electrolyte and the electronic conductor, since this lengthens the TPB. In order to achieve effective ion-conduction pathways from the dense electrolyte, into the composite electrode, high-temperature sintering is normally required. Unfortunately, this high-temperature sintering step limits the choice of materials that can be used. For example, Sr-doped $LaMnO_3$ (LSM) is the most commonly used cathode material for SOFC based on yttria-stabilized zirconia (YSZ) electrolytes, even though cathodes with Sr-doped $LaFeO_3$ (LSF) and Sr-doped $LaCoO_3$ (LSCo) exhibit much better performance, primarily because LSM and YSZ are less reactive, allowing higher sintering temperatures during fabrication.

Recently, we have developed a novel approach for preparing composite electrodes that allows separate temperatures to be used for sintering the electrolyte material and for fabricating the electronic components.[3-6] The approach involves preparing highly porous YSZ, sintered together with the electrolyte,[7] then adding the other components in the electrode by impregnation. The porous YSZ structures are prepared by tape-casting methods, with pore formers added to the green tapes. They are fired to high temperatures (~1800 K) together with the electrolyte, insuring high ionic conduction between the dense electrolyte and the electrode and within the porous YSZ itself. Because the electroactive components of the electrodes are added after the high-temperature sintering of the YSZ has been completed, this second component need not be exposed to such high temperatures.

In this paper, we describe the methods used in the preparation of SOFC by impregnation and then present results that demonstrate that anodes and cathodes with interesting properties can be prepared in this manner.

FABRICATION PROCEDURES

Electrolyte Matrix

The first step in preparing electrodes by impregnation involves synthesizing a porous matrix of the electrolyte material together with the dense electrolyte. While a porous-dense multilayer structure can be produced by a number of methods, we have found tape casting to be the simplest. After ball milling the YSZ powder with a dispersant to break up agglomerates, it is mixed with binders to produce a slurry. This slurry is used directly for tape casting the dense electrolyte layer or mixed with various, pyrolyzable pore formers to produce a slurry for casting the porous layers. We have used primarily graphite, polymethyl-methacrylate, or polystyrene as pore formers, depending on the size and shape of pores that are desired.[8] After producing green tapes, with and without pore formers, the tapes are laminated to produce the multilayer, green tape. For added strength, structural layers, YSZ tapes with holes, can be laminated onto the multilayer green tape. Finally, the tapes are fired to 1823 K for 4 h to produce the YSZ matrix.

An example of the kind of YSZ structure that can be prepared in this way is shown in Fig. 1. The sample in this SEM micrograph is pure YSZ, with two porous layers separated by a 25-μm dense layer in the middle. The porosity of the YSZ on both sides of this layer is approximately 65% but the bottom layer contains both large and small pores due to the use of both graphite and polystyrene pore formers. We have found that the smaller pores are essential for enhancing electrode performance,[9] probably due to providing a longer TPB, while the larger pores are likely to provide enhanced diffusion of the reactant gases.

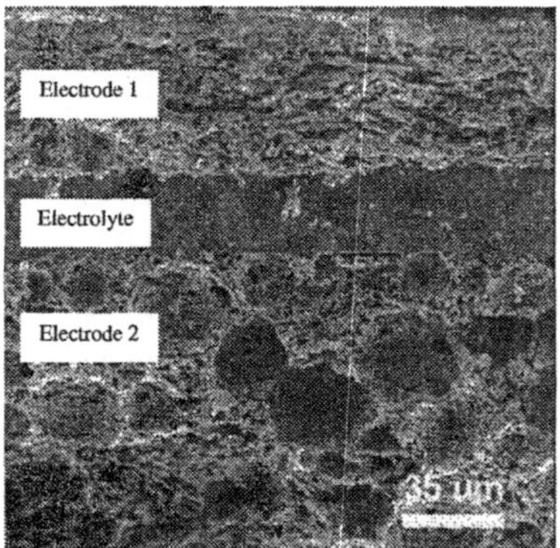

Figure 1. An SEM micrograph of a dense YSZ layer separated by two porous YSZ layers that could be used to prepare SOFC electrodes.

Anode Preparation

Both the anode and cathode composites can be fabricated using impregnation of the porous YSZ matrices with aqueous solutions, as described in other papers. Considering first the anodes, a wide range of compositions can be achieved by impregnation. In most of our work at Penn, the focus has been on the preparation of Cu-ceria-YSZ electrodes for the direct oxidation of hydrocarbons.[4,5] For this application, a typical anode composition is 15 wt% CeO_2 and 30 wt% Cu prepared by incorporating aqueous solutions of $Ce(NO_3)_3{\cdot}6H_2O$ and $Cu(NO_3)_2{\cdot}6H_2O$ into the structure. Achieving the final desired composition requires multiple impregnation steps, with intermediate heating to 723 K in air in order to decompose the nitrates.

Capillary forces can lead to preferential precipitation of the salts at the surface of the porous matrix during the drying process.[10] Having additional Cu at the surface is not necessarily bad, since excess Cu at the surface of the electrode can assist in current collection; however, preferential deposition can be prevented by adding urea to the aqueous solution that is used during impregnation.[10] As the solution is heated, the urea reacts with water to form ammonia, dramatically increasing the pH and causing the salts to precipitate as hydroxides before the water has been removed.

In the Cu-ceria-YSZ composite anodes, it appears that the catalytic activity is supplied solely by the ceria, with Cu providing only electronic conductivity. The evidence for this is as follows. First, Cu-YSZ anodes that do not contain ceria exhibit very low performance, even though they are stable in hydrocarbon fuels.[3,11] Second, substitution of Cu with Au has essentially no effect on anode performance.[12] Since Au is usually thought to be catalytically inert, it seems unlikely that Cu-based anodes and Au-based anodes would perform in a similar manner if Cu was providing catalytic activity. Finally, the performance of cells with anodes containing only ceria and YSZ can approach that of cells which also have Cu if they are exposed to hydrocarbons at elevated temperatures in order to form an electronically conductive layer of tar within the anode pores.[13,14]

An important implication of the fact that Cu is acting strictly as an electronic conductor within the Cu-ceria-YSZ anodes is that these anodes are very sulfur tolerant.[15] Equilibrium considerations show that sulfur poisoning of Ni-based anodes cannot be due to formation of bulk Ni sulfides, since these should not be stable except at high sulfur concentrations. Rather, sulfur binds strongly to the Ni surface and forms surface sulfide layers at low sulfur concentrations and these suppress the catalytic activity of the Ni. In the case of the Cu-ceria-YSZ anodes, sulfur poisoning results from the formation of Ce_2O_2S, which requires much higher sulfur concentrations in the fuel.

The impregnation method can also be used to prepare anodes based on Cu-Ni alloys and Cu-Co mixtures.[16-18] Anodes based on Cu-Co mixtures are particularly intriguing in that they exhibit some of the high catalytic activity and good thermal stability of Co but are reasonably tolerant against carbon deposition in the presence of hydrocarbon fuels. Indeed, it has been shown that an anode prepared from a 50:50 mixture of Cu and Co was stable in dry methane at 1073 K for 500 h.[18] Because Cu and Co do not form alloys, it appears that the carbon tolerance of the Cu-Co mixtures is a result of Cu migrating to the surface of the metal phase. Thermal stability of the metal phase is likely attributable to the Co "core". Because Co has a slight solubility in Cu, the enhanced catalytic activity of the Cu-Co anodes may result from isolated Co atoms at the Cu-rich surface. These isolated atoms would be expected to assist in the electrochemical reaction but would not be active for the growth of carbon fibers.

Cathode Preparation

Composite cathodes based on LSM, LSF, and LSCo can be made using methods very similar to that used in preparing anodes. For example, LSF-YSZ composites can be fabricated by impregnation of the YSZ matrix with aqueous solutions containing $La(NO_3)_3 \cdot 6H_2O$, $Sr(NO_3)_2$, and $Fe(NO_3)_3 \cdot 9H_2O$ at a molar ratio of La:Sr:Fe = 0.8:0.2:1, following by calcination above 1073 K.[19] To produce single-phase LSF, it is necessary to ensure intimate mixing of the cations in the solid phase that precipitates in the matrix. This can be achieved through the addition of citric acid to the solution in an amount equal to that of the metal ions. LSCo-YSZ composites can be prepared in a similar manner, but the addition of citric acid is not required.[9]

Composite cathodes can also be prepared through the impregnation of oxide nanoparticles, a procedure that has two advantages over impregnation of salt solutions. First, the oxide nanoparticles can have an intimate cation mixture, or even be in the perovskite structure, before addition to the porous matrix. This avoids the need for additives like citric acid to promote interactions. Second, nanoparticle solutions can be formed in high concentrations, so that it should be possible to form the impregnated electrode in a single step. Fig. 2 shows SEM micrographs of porous YSZ with LSF incorporated to a level of 30 wt%, near the dense electrolyte interface, prepared both by impregnation of the aqueous salts and by impregnation of nanoparticles. The appearance is similar in both cases, with the LSF particles existing as small, ~0.1-μm particles, on the surface of the YSZ. In both cases, X-ray diffraction measurements showed the presence of only two phases, those of YSZ and LSF.

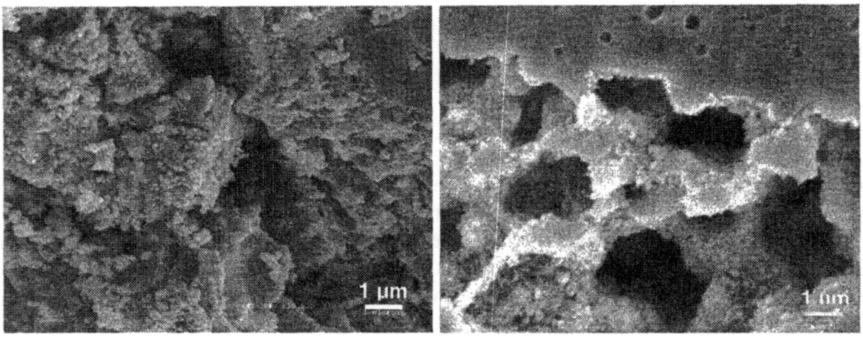

Figure 2. SEM micrographs of LSF-YSZ composites prepared with 35-wt% LSF. On the left, LSF was added using the aqueous nitrate salts, while the composite on the right was formed by impregnating LSF nanoparticles. Dense YSZ is shown at the top of this micrograph, demonstrating that the particles could reach the region near the electrolyte.

In addition to the fact that the impregnated cathodes can be made from essentially any oxide, this procedure allows the preparation of high-performance, cathode-supported cells.[19] Cathode-supported cells could have significant advantages at high fuel utilization, since they would allow the anodes to be very thin, avoiding diffusional limitations that could decrease performance.[20] While the in-plane conductivity of composites formed by conducting oxides will not likely approach that of metals, the through-plane conductivity can be high enough so that the ohmic resistance of the cathode will be negligible. For example, even if the conductivity of the electrode is only 1 S/cm, an electrode of 0.05-cm thickness will contribute only 0.05 $\Omega \cdot cm^2$ towards the cell impedance.

Finally, it should be noted that preparation of electrodes by impregnation can result in fewer calcination steps during SOFC fabrication. Since the porous YSZ layers are calcined together with the electrolyte, there may be no need for additional firing of the electrode after addition of the active components.

PROPERTIES CHARACTERIZATION

Composite electrodes prepared by most conventional methods are essentially random. One demonstration of this is found in conductivity measurements of Ni-YSZ composites made with varying Ni loadings,[21] which show a sharp increase in the conductivity of the composite at ~30-vol% Ni. This concentration corresponds to the percolation threshold for random media, and is the amount of metal required to form a completely interconnected metal phase in the composite. By contrast, the composites prepared by impregnation are not random since the conducting phase essentially coats the electrolyte phase. This has an important effect on a number of composite properties.

First, the electrical conductivities of the composites can be reasonably high at low loadings of the electronically conductive phase. For example, the conductivity of a Cu-YSZ composite was reported to reach 10 S/cm at a Cu loading of only 5 vol%.[14] A second example is shown in Table 1, which reports the electrical conductivities, measured in air at 973 K, of selected LSF-YSZ and LSCo-YSZ composites prepared from a 60% porous, YSZ matrix. As discussed in a previous paper,[9] the conductivity of the porous YSZ, ~10^{-3} S/cm, was 10 times lower than the accepted value for YSZ due to the sample porosity. The conductivity increased steadily with LSF loading to 1.3 S/cm after addition of 40-wt% LSF. In the LSCo-YSZ composites, the conductivity was 0.2 S/cm with only 10-wt% LSCo, increasing to 30 S/cm at 40-wt% LSF. Random mixtures of conducting and insulating materials should not exhibit the gradual increase in conductivity that was observed with these composites and the increase in conductivity should have been minimal below the percolation threshold (30 vol%), which would occur at weight loadings near 50% for these materials. While the absolute conductivities of the composites are still much lower than those of pure LSF or LSCo, they are comparable to those measured for a 50-wt% LSM-YSZ composite, 8.5 S/cm, prepared by conventional methods.[6]

Table 2. Electrical conductivities at 973 K in air of LSF-YSZ and LSCo-YSZ composites prepared by impregnation methods.

LSF or LSCo Weight fraction in YSZ	0%	10%	30%	40%
σ (S/cm) of LSF-YSZ composites	10^{-3}	0.05	0.8	1.3
σ (S/cm) of LSCo-YSZ composites	10^{-3}	0.2	7	30

The CTE of several LSCo-YSZ composites were measured between room temperature and 1073 K, with the data shown in Table 2. The CTE for the porous YSZ was within experimental error of the literature value for pure YSZ, 10.5×10^{-6}/K. The CTE increased slightly with increased loadings of LSCo, to 12.6×10^{-6}/K at 55-wt% LSCo. However, this increase in CTE is much less than would be expected if the CTE of the composite were simply a weight-averaged value of YSZ and LSCo, which has a CTE of 23×10^{-6}/K. Again, it appears that the CTE of these composites is dominated by the YSZ backbone, giving the composites properties that are quite different from that of normal, random composites.

Table 2. CTE values of the LSCo-YSZ composites prepared by impregnation.

LSCo weight fraction in YSZ	0%	35%	45%	55%
CTE (10^{-6} /K)	10.3	11.7	12.6	12.6

The electrochemical properties of impregnated electrodes can also be quite good. For example, Fig. 3 shows impedance data at 973, 1023, and 1073 K in air for symmetric cells made with 35-wt% LSCo in YSZ. Prior to these measurements, the LSCo-YSZ sample was heated in air to 1073 K for 3 h. The results in the Cole-Cole plots have been multiplied by 0.5 to account for the two electrodes. The ohmic resistance, which was approximately equal to that expected for the electrolyte, has been removed. The ASR decreased from 0.03 $\Omega\cdot cm^2$ at 973 K to less than 0.015 $\Omega\cdot cm^2$ at 1073 K. At the lower temperatures, there appear to be two arcs in the impedance spectra, with characteristic frequencies of 10 Hz and 500 Hz. The two arcs have different temperature dependences, with the 500-Hz feature essentially disappearing at 1073 K. Therefore, we suggesting that the rate-limiting step on the LSCo-YSZ composite changes with temperature.

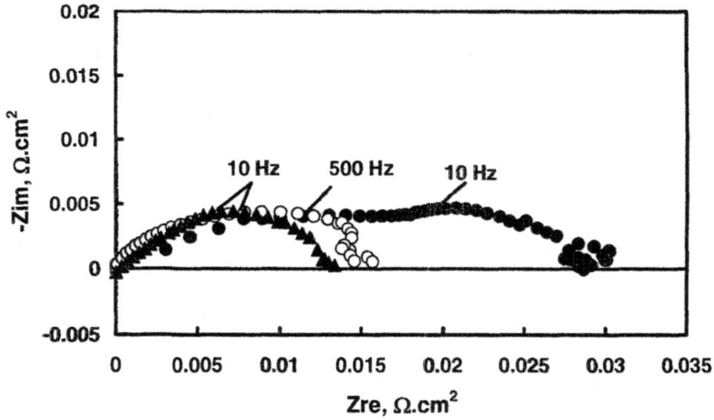

Figure 3. Impedance spectra, measured in air using symmetric cells, for 35-wt% LSCo in YSZ. The measurements were performed at: ● 973 K, ○ 1023 K, and ▲ 1073 K.

That impregnated electrodes can perform well under fuel cell conditions is shown in Fig. 4. The cell used in this data had a cathode that was 35-wt% LSCo in YSZ, a 40-μm thick YSZ electrolyte, and an anode containing 15-wt% CeO_2 and 30-wt% Co. The V-I polarization curves in humidified H_2 (3% H_2O) are linear over the entire range of currents tested, with no evidence for polarization activation at open circuit. The cell exhibited maximum power densities of 0.48, 0.68, and 0.88 W/cm^2 at temperatures of 973 K, 1023 K, and 1073 K, respectively. This performance level is comparable to that of the very best YSZ-based SOFC, given the relatively thick electrolyte in this cell.

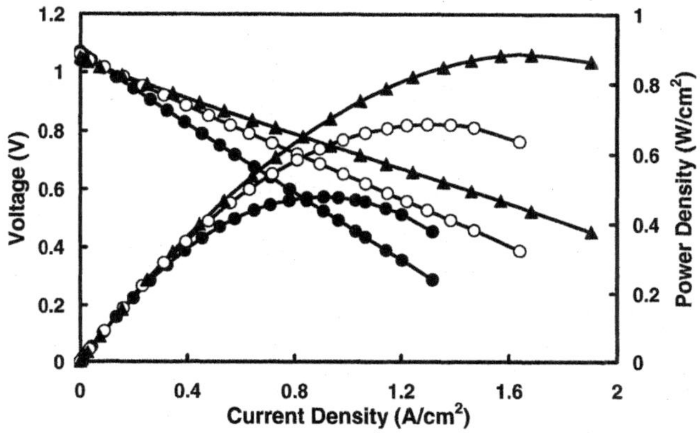

Figure 4. Performance curves measured in humidified H_2 for a cell with both an impregnated, Co-ceria-YSZ anode and an impregnated LSCo-YSZ cathode. The electrolyte was YSZ, 40-μm thick. The temperatures for these measurements were as follows: ● 973 K, ○ 1023 K, and ▲ 1073 K.

CONCLUSIONS

We have developed a novel method for fabricating composite SOFC anodes and cathodes that involves impregnating a porous YSZ matrix with electroactive components. Because the method allows different sintering temperatures for each phase in the composite, a much wider range of electrode compositions is possible. Furthermore, the impregnated electrodes have properties, CTE and conductivity, that differ significantly from that of traditional electrodes because the electroactive phase is not random, but rather coats the walls of the YSZ phase. Finally, it was shown that the electrochemical performance of the electrodes can be excellent.

ACKNOWLEDGEMENT

We acknowledge support from the Office of Naval Research.

REFERENCES

[1]S. McIntosh and R. J. Gorte, "Direct hydrocarbon solid oxide fuel cells," *Chem. Rev.*, **104**, 4845-65 (2004).

[2]M. L. Toebes, J. H. Bitter, A. J. van Dillen, and K. P. de Jong, "Impact of the structure and reactivity of nickel particles on the catalytic growth of carbon nanofibers," *Catal. Today*, **76**, 33-42 (2002).

[3]R. Craciun, S. Park, R. J. Gorte, J. M. Vohs, C. Wang, and W. L. Worrell, "A novel method for preparing anode cermets for solid oxide fuel cells," *J. Electrochem. Soc.*, **146**, 4019-22 (1999).

[4]R. J. Gorte, S. Park, J. M. Vohs, and C. H. Wang, "Anodes for direct oxidation of dry hydrocarbons in a solid-oxide fuel cell," *Adv. Matls.*, **12**, 1465-69 (2000).

[5]S. D. Park, J. M. Vohs, and R. J. Gorte, "Direct oxidation of hydrocarbons in a solid-oxide fuel cell," *Nature*, **404**, 265-7 (2000).

[6]H. P. He, Y. Y. Huang, J. Regal, M. Boaro, J. M. Vohs, and R. J. Gorte, "Low-temperature fabrication of oxide composites for solid-oxide fuel cells," *J. Am. Ceram. Soc.*, **87**, 331-6 (2004).

[7]S. Park, R. J. Gorte, and J. M. Vohs, "Tape cast solid oxide fuel cells for the direct oxidation of hydrocarbons," *J. Electrochem. Soc.*, **148**, A443-7 (2001).

[8]M. Boaro, J. M. Vohs, and R. J. Gorte, "Synthesis of highly porous yttria-stabilized zirconia by tape-casting methods," *J. Am. Ceram. Soc.*, **86**, 395-400 (2003).

[9]Y. Y. Huang, K. Ahn, J. M. Vohs, and R. J. Gorte, "Characterization of Sr-doped $LaCoO_3$-YSZ composites prepared by impregnation methods," *J. Electrochem. Soc.*, **151**, A1592-97 (2004).

[10]S. Jung, C. Lu, H. He, K. Ahn, R. J. Gorte, and J. M. Vohs, "Influence of Composition and Cu Impregnation Method on the Performance of Cu/CeO2/YSZ SOFC Anodes", *J. Power Sources*, in press.

[11]S. McIntosh, J. M. Vohs, and R. J. Gorte, "An examination of lanthanide additives on the performance of Cu-YSZ cermet anodes," *Electrochim. Acta*, **47**, 3815-21 (2002).

[12]C. Lu, W. L. Worrell, J. M. Vohs, and R. J. Gorte, "A comparison of Cu-ceria-SDC and Au-ceria-SDC composites for SOFC anodes", *J. Electrochem. Soc.*, **150**, A1357-59 (2003).

[13]S. McIntosh, J. M. Vohs, and R. J. Gorte, "Role of hydrocarbon deposits in the enhanced performance of direct-oxidation SOFCs,", *J. Electrochem. Soc.*, **150**, A470-76 (2003).

[14]S. McIntosh, H. P. He, S.-I. Lee, O. Costa-Nunes, V. V. Krishnan, J. M. Vohs, and R. J. Gorte, "An examination of carbonaceous deposits in direct-utilization SOFC anodes," *J. Electrochem. Soc*, **151**, A604-08 (2004).

[15]H. He, R. J. Gorte, and J. M. Vohs, "Highly Sulfur Tolerant Cu-Ceria Anodes for SOFCs", *Electrochem. & Sol.State Lett.*, **8**, A279-80 (2005).

[16]H. Kim, C. Lu, W. L. Worrell, J. M. Vohs, and R. J. Gorte, "Cu-Ni cermet anodes for direct oxidation of methane in solid-oxide fuel cells," *J. Electrochem. Soc.*, **149**, A247-50 (2002).

[17]S.-I. Lee, "A study of SOFC anodes based on Cu-Ni and Cu-Co bimetallics in CeO2-YSZ," *J. Electrochem. Soc.*, **151**, A1319-23 (2004).

[18]S.-I. Lee, K. Ahn, J. M. Vohs, and R. J.Gorte, "Cu-Co Bimetallic Anodes for direct utilization of methane in SOFCs", *Electrochem. & Sol. State Lett.*, **8**, A48-51 (2005).

[19]Y. Y. Huang, J. M. Vohs, and R. J. Gorte, "Fabrication of Sr-doped $LaFeO_3$YSZ composite cathodes,", *J. Electrochem. Soc.*, **151**, A646-51 (2004).

[20]O. Costa-Nunes, J. M. Vohs, and R. J. Gorte, "A study of direct-conversion SOFC with n-butane at higher fuel utilization," *J. Electrochem. Soc.*, **150**, A858-63 (2003).

[21]D. W. Dees, T. D. Claar, T. E. Easler, D. C. Fee, and F. C. Mrazek, "Conductivity of porous Ni/ZrO_2-Y_2O_3 Cermets," *J. Electrochem. Soc.*, **134**, 2141-46 (1987).

INVESTIGATION OF $Nd_{0.6}Sr_{0.4}Co_{1-y}M_yO_{3-\delta}$ (M = Fe AND Mn) AS CATHODE MATERIALS FOR INTERMEDIATE TEMPERATURE SOLID OXIDE FUEL CELLS

K. T. Lee and A. Manthiram
Materials Science and Engineering Program
University of Texas at Austin
Austin, TX 78712

ABSTRACT

$Nd_{0.6}Sr_{0.4}Co_{1-y}M_yO_{3-\delta}$ (M = Fe and Mn) perovskite oxides have been investigated as cathodes for intermediate temperature solid oxide fuel cells (SOFC) with $La_{0.8}Sr_{0.2}Ga_{0.8}Mg_{0.2}O_{2.8}$ electrolyte. The crystal chemistry, electrical conductivity, thermal expansion behavior, and electrochemical performance of the cathodes are presented. The oxygen non-stoichiometry and electrical conductivity decrease with increasing Fe and Mn contents. Both the Fe- and Mn-doped systems exhibit a metal to semiconductor transition at $y \approx 0.2$, but the Mn-doped system exhibits higher activation energy than the Fe-doped system. The thermal expansion coefficient (TEC) decreases with increasing Fe and Mn contents due to a suppression of the low-spin to high-spin transition associated with the Co^{3+} ions and a stronger M-O bonding. The electrocatalytic activity decreases with increasing Fe and Mn contents, but the decrease is rapid with the Mn-doped system due to a faster decrease in the oxide ion and electronic conductivities.

INTRODUCTION

Solid oxide fuel cells (SOFCs) offer the advantage of directly using the hydrocarbon fuels without external reforming as well as high energy conversion efficiency. However, the high operating temperature of ~ 1000 °C associated with the conventional yttria-stabilized zirconia (YSZ) electrolytes leads to thermal expansion mismatch, interfacial reaction among the components, and a limited choice of interconnect and cathode materials.[1,2] These difficulties could be minimized by lowering the operating temperature to an intermediate range of 500-800 °C, but the lower temperature leads to large over-potentials due to slow oxygen reduction kinetics at the cathode. Although the Sr-doped $LaMnO_3$ cathode is used with the YSZ electrolyte at ~ 1000 °C, its low oxide ion conductivity prevents it from being used for intermediate temperature SOFCs.[3,4] In this regard, Sr-doped $LaCoO_3$ is appealing due to its high electronic and ionic conductivities, but it experiences high thermal expansion.[5,6] Replacement of La by other smaller lanthanides could lower the thermal expansion due to the decreased ionicity of the Ln-O bond. We showed recently that Sr-doped $NdCoO_3$ offers a combination of acceptable electronic conductivity and thermal expansion.[7] We present here the crystal chemistry, electrical conductivity, thermal expansion, and electrochemical performance of $Nd_{0.6}Sr_{0.4}Co_{1-y}M_yO_{3-\delta}$ (M = Fe and Mn) perovskite oxide cathodes.

EXPERIMENTAL

The $Nd_{0.6}Sr_{0.4}Co_{1-y}M_yO_{3-\delta}$ (M = Fe and Mn) cathode materials were synthesized by firing required amounts of Nd_2O_3, $SrCO_3$, Co_3O_4, and Fe_2O_3 or Mn_2O_3 in air at 900 °C for 12 h, followed by regrinding, pressing into pellets, and sintering for 24 h at 1200 °C for the Fe-doped system and 1300 °C for the Mn-doped system. The sintering was repeated for another 24 h after regrinding and repelletizing in order to improve the product homogeneity. The $La_{0.8}Sr_{0.2}Ga_{0.8}Mg_{0.2}O_{2.8}$ (LSGM) electrolyte was prepared by firing required amounts of La_2O_3, $SrCO_3$, Ga_2O_3, and MgO at 1100 °C for 5 h, followed by pelletizing and sintering at 1500 °C for 10 h. The NiO-$Ce_{0.9}Gd_{0.1}O_{1.95}$ (GDC) cermet (Ni:GDC = 70:30 vol %) anode was prepared by the glycine-nitrate combustion method.[8]

Structural characterizations were carried out by X-ray diffraction using the Rietveld method. Thermogravimetric analysis (TGA) was carried out up to 1000 °C in air with heating and cooling rates of, respectively, 2 and 10 °C/min. TECs of the sintered samples were measured up to 700 °C with a heating/cooling rate of 10 °C/min. Electrical conductivity measurements were carried out with a four-probe dc method in the temperature range of 200-900 °C. All the samples used for thermal expansion and conductivity measurements had densities of > 95% of the theoretical density values. Electrochemical performances were evaluated with single cells at 800 °C. The $Nd_{0.6}Sr_{0.4}Co_{1-y}M_yO_{3-\delta}$ (M = Fe and Mn) cathodes and NiO-GDC cermet anode were prepared by screen printing onto the LSGM electrolyte pellet, followed by firing for 3 h at 1000 and 1100 °C, respectively, for the Fe-doped and Mn-doped cathodes and 1200 °C for the anode. The geometrical area of the electrode was 0.25 cm^2. Humidified H_2 (~3% H_2O at 30 °C) and air were supplied as fuel and oxidant, respectively, at a rate of 100 cm^3/min. I-V measurements were carried out using a three electrode configuration, which allowed for separating and monitoring cathode over-potentials during operation. Pt paste was used as the reference electrode.

RESULTS AND DISCUSSION

All the $Nd_{0.6}Sr_{0.4}Co_{1-y}M_yO_{3-\delta}$ (M = Fe and Mn) samples were found to be single phase and the X-ray diffraction patterns could be indexed with the orthorhombic $GdFeO_3$ type perovskite structure for $0 \leq y \leq 0.5$. The lattice parameters and BET surface area of the samples are summarized in Table I. The lattice parameters and the lattice volume increase with increasing y due to the replacement of the smaller Co^{3+}/Co^{4+} (average ionic radius, r = 0.538 Å) ions by the larger Fe^{3+}/Fe^{4+} (r = 0.615 Å) or Mn^{3+}/Mn^{4+} (r = 0.589 Å) ions.

Table I. Crystal chemistry data and BET surface area of $Nd_{0.6}Sr_{0.4}Co_{1-y}M_yO_{3-\delta}$ (M = Fe and Mn)

M	y	a (Å)	b (Å)	c (Å)	Lattice volume (Å^3)	BET surface area (m^2/g)
Fe	0.0	5.3626(5)	5.4111(4)	7.5859(8)	220.1	4.8
	0.1	5.3693(5)	5.4177(5)	7.5964(7)	221.0	4.0
	0.2	5.3783(5)	5.4256(4)	7.6093(7)	222.0	5.0
	0.3	5.3867(6)	5.4317(5)	7.6200(8)	223.0	4.9
	0.4	5.3974(6)	5.4385(5)	7.6352(7)	224.1	5.2
	0.5	5.4113(6)	5.4485(5)	7.6547(8)	225.7	4.2
Mn	0.0	5.3626(5)	5.4111(4)	7.5859(8)	220.1	4.2
	0.1	5.3632(4)	5.4127(4)	7.5872(6)	220.3	4.3
	0.2	5.3677(4)	5.4163(4)	7.5922(6)	220.7	3.1
	0.3	5.3691(4)	5.4169(4)	7.5955(6)	220.9	4.1
	0.4	5.3757(4)	5.4219(3)	7.6042(5)	221.6	3.6
	0.5	5.3839(4)	5.4287(3)	7.6139(5)	222.5	3.8

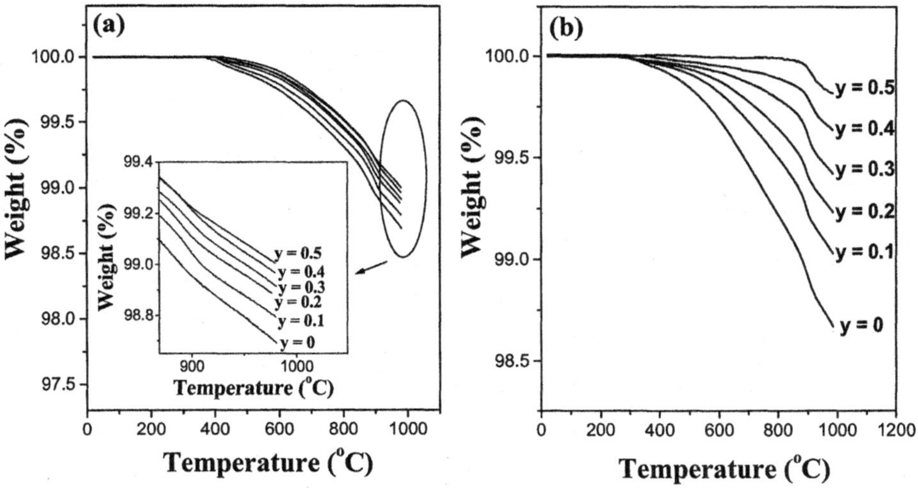

Figure 1. TGA plots of $Nd_{0.6}Sr_{0.4}Co_{1-y}M_yO_{3-\delta}$ ($0 \le y \le 0.5$) recorded in air with a heating rate of 2 °C/min: (a) M = Fe and (b) M = Mn.

The TGA plots of the $Nd_{0.6}Sr_{0.4}Co_{1-y}M_yO_{3-\delta}$ samples recorded in air are shown in Figure 1. The amount of oxygen loss occurring at high temperatures decreases with increasing Fe or Mn content. Tai *et al.*[9] and Kozhevnikov *et a*l.[10] reported a similar trend for the analogous $La_{1-x}Sr_xCo_{1-y}Fe_yO_3$ and $La_{1-x}Sr_xCo_{1-z}Mn_zO_{3-\delta}$ systems respectively. The decrease in the amount of oxygen loss with Fe- and Mn- doping could be attributed to the increase in the binding energy of the oxide ions in the lattice. This is in agreement with the report by Steele[11] that perovskite oxides with Fe^{3+} and Mn^{3+} ions are thermally more stable than those with Co^{3+} ions due to the stronger Fe-O and Mn-O bonds compared to the Co-O bond. It is also in accordance with the finding that the oxygen non-stoichiometry δ in $La_{1-x}Sr_xFeO_{3-\delta}$ and $La_{1-x}Sr_xMnO_{3-\delta}$ is almost fixed at $\delta \approx 0.5x$ over a wide range of pO_2, while that in $La_{1-x}Sr_xCoO_{3-\delta}$ increases with decreasing pO_2.[12,13]

The temperature dependence of the electrical conductivity of the $Nd_{0.6}Sr_{0.4}Co_{1-y}M_yO_{3-\delta}$ (M = Fe and Mn) samples is shown in Figure 2. The faster decrease in conductivity at higher temperatures could be due to the oxygen loss from the lattice at higher temperatures as indicated by the TGA data (Figure 1) and the consequent decrease in carrier concentration. Additionally, the increasing oxygen vacancy concentration can perturb the periodic potential of the (Co,M)-O-(Co,M) network and lead to a trapping of the carriers.

At a given temperature, the electrical conductivity decreases with increasing Fe and Mn contents. The $Nd_{0.6}Sr_{0.4}Co_{1-y}M_yO_{3-\delta}$ (M = Fe and Mn) samples are mixed hopping conductors with more than one type of transition metal ions. It has been reported that the electronic charge compensation occurs preferentially by the formation of Fe^{4+} and Mn^{4+} rather than Co^{4+} on substituting Sr^{2+} for lanthanide ions, respectively, in the $La_{0.6}Sr_{0.4}Co_{1-y}Fe_yO_{3-\delta}$ and $Pr_{1-x}Sr_xCo_{1-y}Mn_yO_{3-\delta}$ systems.[14,15] Thus, the preferential formation of Fe^{4+} and Mn^{4+} compared to Co^{4+} and the decreased covalency of the Fe^{4+}-O and Mn^{4+}-O bonds compared to the Co^{4+}-O bonds result in an increase in the electron localization and a decrease in the electrical conductivity with increasing Fe or Mn content.

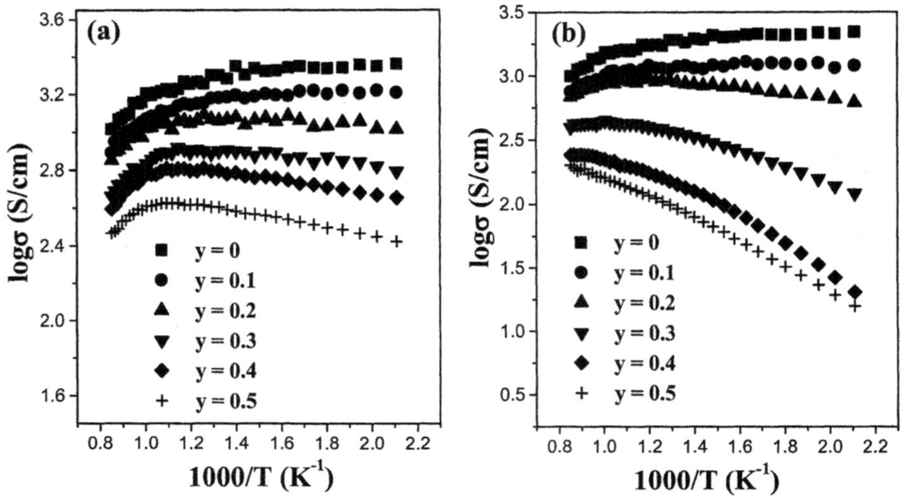

Figure 2. Variations of the electrical conductivity measured in air of (a) $Nd_{0.6}Sr_{0.4}Co_{1-y}Fe_yO_{3-\delta}$ and (b) $Nd_{0.6}Sr_{0.4}Co_{1-y}Mn_yO_{3-\delta}$ with temperature for various value of y.

The $Nd_{0.6}Sr_{0.4}Co_{1-y}M_yO_{3-\delta}$ (M = Fe and Mn) system exhibits a metal to semiconductor transition around y = 0.2 as seen in Figure 2. The metal to semiconductor transition in $Nd_{0.6}Sr_{0.4}Co_{1-y}M_yO_{3-\delta}$ (M = Fe and Mn) with Fe and Mn doping could be explained by considering the changes in the structural parameters. The end member $Nd_{0.6}Sr_{0.4}CoO_{3-\delta}$ is a metallic conductor with sufficient bandwidth and little or no charge transfer gap between the $Co^{3+/4+}$:3d and O^{2-}:2p bands.[7] The substitution of a larger Fe^{3+} or Mn^{3+} ion for Co^{3+} leads to an increase in the average ionic radius of the B site cations in the ABO_3 perovskite structure, resulting in a decrease in the tolerance factor and an increase in the bending of the O-(Co,M)-O bonds with bond angles of < 180°. The increase in the bending of the O-(Co,M)-O bonds, causing a decrease in the bandwidth and a consequent opening of the charge transfer gap, could be envisioned to cause a metal to semiconductor transition in the $Nd_{0.6}Sr_{0.4}Co_{1-y}M_yO_{3-\delta}$ (M = Fe and Mn) system with increasing Fe or Mn doping.

The activation energies obtained from the Arrhenius plots of ln(σT) versus 1/T are given in Table II. The activation energy increases with increasing Fe and Mn contents. The temperature dependence of the conductivity for a semiconductor can be described by the small polaron hopping mechanism and the activation energy for hopping is strongly related to polaron binding energy and bandwidth. The substitution of a larger Fe^{3+} and Mn^{3+} for Co^{3+} and the preferential formation of Fe^{4+} and Mn^{4+} compared to Co^{4+} result in a bending of the O-(Co,M)-O bonds and deep localization of charge carriers with a decrease in bandwidth. Thus, the decrease in the bandwidth and the increase in polaron binding energy lead to an increase in the activation energy with increasing Fe and Mn contents. For a given degree of doping, the Mn-doped system shows higher activation energies than the Fe-doped system, which implies a less covalency and larger charge transfer gap for the former compared to the latter.

Table II. Activation energies (E_a) for electrical conduction in $Nd_{0.6}Sr_{0.4}Co_{1-y}M_yO_{3-\delta}$ (M = Fe and Mn) at 200-700 °C in air

y	E_a for M = Fe (eV)	E_a for M = Mn (eV)
0.2	0.060	0.087
0.3	0.070	0.159
0.4	0.084	0.237
0.5	0.094	0.238

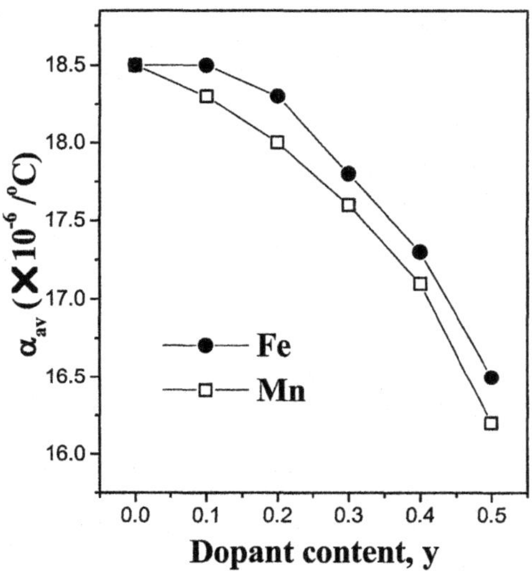

Figure 3. Variations of the average thermal expansion coefficients (50-700 °C) of $Nd_{0.6}Sr_{0.4}Co_{1-y}M_yO_{3-\delta}$ (M = Fe and Mn) with dopant concentration y.

The average thermal expansion coefficient values of $Nd_{0.6}Sr_{0.4}Co_{1-y}M_yO_{3-\delta}$ (M = Fe and Mn) at 50-700 °C in air are plotted as a function of dopant concentration y in Figure 3. The TEC values decrease with increasing Fe or Mn content y. The low spin ($t_{2g}^6e_g^0$) to high spin ($t_{2g}^4e_g^2$) transition of Co^{3+} ions normally occurring with increasing temperature leads to a larger TEC in $Ln_{1-x}Sr_xCoO_{3-\delta}$.[16] It has also been reported that high spin Co^{3+} ($t_{2g}^4e_g^2$) ions could be stabilized at lower temperatures relative to the low spin Co^{3+} ($t_{2g}^6e_g^0$) ions by the introduction of Fe^{3+} into the $LaCo_{1-x}Fe_xO_3$ system.[17] Therefore, a decrease in the Co^{3+} content and a suppression of the spin state transitions due to Fe^{3+} and Mn^{3+} doping could cause a decrease in the TEC value with increasing y. Furthermore, TEC is inversely proportional to the bonding energy between the ions in the lattice.[18] Thus, the greater decrease in TEC with the Mn doping compared to the Fe doping could be attributed to the stronger Mn-O bond compared to the Fe-O bond.

The X-ray powder diffraction patterns of $Nd_{0.6}Sr_{0.4}Co_{1-y}M_yO_{3-\delta}$ (M = Fe and Mn) and LSGM mixtures after heating for 3 h at 1000 °C for the Fe-doped cathode and 1100 °C for the Mn-doped cathode are shown in Figure 4. No reaction product is observed for all compositions.

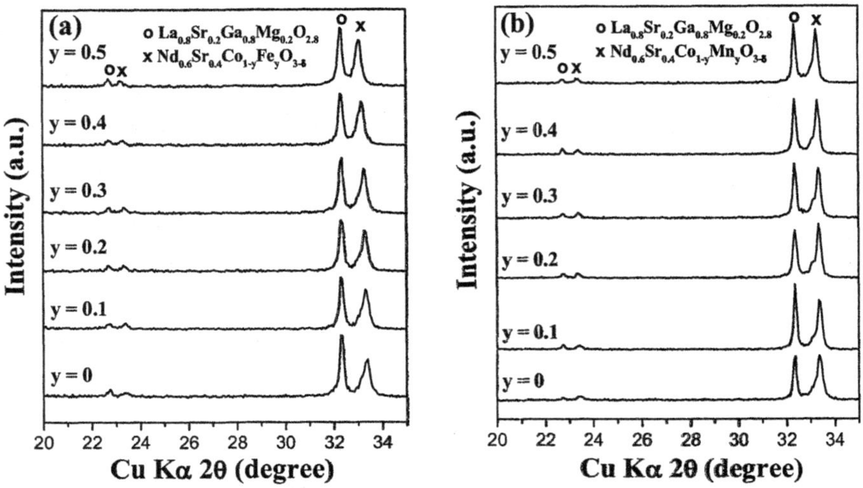

Figure 4. X-ray powder diffraction patterns recorded after heating the $Nd_{0.6}Sr_{0.4}Co_{1-y}M_yO_{3-\delta}$ cathode and the LSGM electrolyte powders for 3 h at (a) 1000 °C for the Fe-doped composition and (b) 1100 °C for the Mn-doped composition.

The BET surface area does not vary significantly with y in $Nd_{0.6}Sr_{0.4}Co_{1-y}M_yO_{3-\delta}$ (M = Fe and Mn) as seen in Table I, and so influence of the geometrical morphology on electrochemical performance in this study could be considered negligible. The variations of the over-potential with current density at 800 °C for the various $Nd_{0.6}Sr_{0.4}Co_{1-y}M_yO_{3-\delta}$ (M = Fe and Mn) cathode compositions are shown in Figure 5. The over-potential increases with increasing Fe and Mn contents y. Cathode over-potential is closely related not only to the charge transfer and the adsorption/dissociation of oxygen, but also to the transport speed of oxide ions through the cathode bulk and across the cathode/electrolyte interface. High oxide ion vacancy concentration in the surface of cathode materials could improve oxygen exchange (the adsorption/dissociation of oxygen molecule). The oxide ion vacancies and electrical conductivity decrease with increasing Fe and Mn doping as seen in Figures 1 and 2. Thus, the decrease in the oxygen exchange, charge transfer, and transport speed of oxide ions caused by a decrease in the oxygen vacancy concentration and electrical conductivity with a substitution of Fe or Mn for Co leads to an increase in over-potential or a decrease in catalytic activity. Additionally, the rapid increase in over-potential for the Mn-doped system compared to the Fe-doped system is due to a faster decrease in the oxide ion and electronic conductivities.

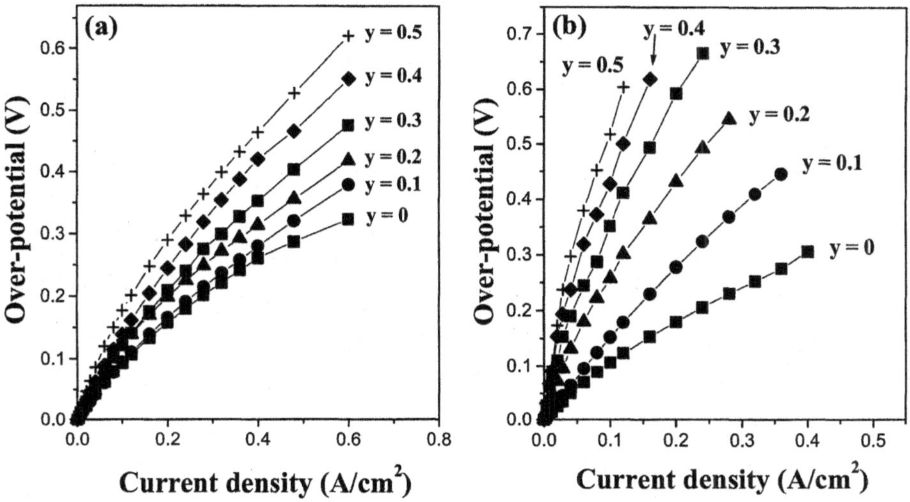

Figure 5. Cathode over-potentials of the $Nd_{0.6}Sr_{0.4}Co_{1-y}M_yO_{3-\delta}$ /LSGM/Ni-GDC single cells at 800 °C for various values of y: (a) M = Fe and (b) M = Mn.

CONCLUSIONS

$Nd_{0.6}Sr_{0.4}Co_{1-y}M_yO_{3-\delta}$ (M = Fe and Mn) perovskite oxides have been synthesized for $0 \leq y \leq 0.5$ and characterized as cathode materials for intermediate temperature solid oxide fuel cells. All the compositions have an orthorhombic structure and the lattice volume increase with increasing Fe and Mn content. The oxygen non-stoichiometry δ decreases with Fe and Mn doping due to stronger Fe-O and Mn-O bonds compared to the Co-O bonds. The electrical conductivity decreases with increasing Fe and Mn content due to the increased carrier localization, and the system exhibits a metal to semiconductor transition around y = 0.2 due to a decrease in the O-(Co,M)-O bond angle below 180° and a consequent decrease in the bandwidth with Fe and Mn doping. The thermal expansion coefficient decreases with increasing Fe and Mn contents due to a suppression of the spin state transitions associated with Co^{3+} and an increase in the bonding energy between the $M^{3+/4+}$ and O^{2-} ions in the lattice. The substitution of $Fe^{3+/4+}$ and $Mn^{3+/4+}$ for $Co^{3+/4+}$ decreases the mobility of the oxide ions as well as the electronic conductivity, which leads to a decrease in electrocatalytic activity. The Fe-doped system exhibits higher electrocatalytic activity but larger TEC than the Mn-doped system. Thus, a trade-off between electrochemical performance and thermal expansion may be necessary to identify optimum cathode compositions.

ACKNOWLEDGEMENT

This work was supported by the Welch Foundation Grant F-1254.

REFERENCES

[1]N.Q. Minh, "Ceramic Fuel Cells," *J. Am. Ceram. Soc.*, **76**, 563-88 (1993).

[2]T. Tsai and S.A. Barnett, "Effect of LSM-YSZ Cathode on Thin-Electrolyte Solid Oxide Fuel Cell Performance," *Solid State Ionics*, **93**, 207-17 (1997).

[3]T. Horita, K. Yamaji, M. Ishikawa, N. Sakai, H. Yokokawa, T. Kawada, and T. Kato, "Active Sites Imaging for Oxygen Reduction at the $La_{0.9}Sr_{0.1}MnO_{3-x}$/Yttria-Sabilized Zirconia Interface by Secondary-Ion Mass Spectrometry," *J. Electrochem. Soc.*, **145**, 3196-202 (1998).

[4]R.A. De Souza and J.A. Kilner, "Oxygen Transport in $La_{1-x}Sr_xMn_{1-y}Co_yO_{3\pm\delta}$ in Perovskites: Part I. Oxygen Tracer Diffusion," *Solid State Ionics*, **106**, 175-87 (1998).

[5]A.N. Petrov, O.F. Kononchuk, A.V. Andreev, V.A. Cherepano, and P. Kofstad, "Crystal Structure, Electrical and Magnetic Properties of $La_{1-x}Sr_xCoO_{3-y}$," *Solid State Ionics*, **80**, 189-99 (1995).

[6]S.B. Adler, "Mechanism and Kinetics of Oxygen Reduction on Porous $La_{1-x}Sr_xCoO_{3-\delta}$ Electrodes," *Solid State Ionics*, **111**, 125-34 (1998).

[7]K.T. Lee and A. Manthiram, "Characterization of $Nd_{1-x}Sr_xCoO_{3-\delta}$ ($0 \leq x \leq 0.5$) Cathode Materials for Intermediate Temperature SOFCs," *J. Electrochem. Soc.*, **152**, A197-A204 (2005).

[8]L.A. Chick, L.R. Pedersen, G.D. Maupin, J.L. Bates, L.E. Thomas, and G.J. Exarhos, "Glycine-Nitrate Combustion Synthesis of Oxide Ceramic Powders," *Materials Letters*, **10**, 6-12 (1990).

[9]L.W. Tai, M.M. Nasrallah, H.U. Anderson, D.M. Sparlin, and S.R. Sehlin, "Structure and Electrical Properties of $La_{1-x}Sr_xCo_{1-y}Fe_yO_3$: Part 1. The System $La_{0.8}Sr_{0.2}Co_{1-y}Fe_yO_3$," *Solid State Ionics*, **76**, 259-71 (1995).

[10]V.L. Kozhevnikov, I.A. Leonidov, E.B. Mitberg, M.V. Patrakeev, A.N. Petrov, and K.R. Poeppelmeier, "Conductivity and Carrier Traps in $La_{1-x}Sr_xCo_{1-z}Mn_zO_{3-\delta}$ ($x = 0.3$; $z = 0$ and 0.25)," *J. Solid State Chem.*, **172**, 296-304 (2003).

[11]B.C.H. Steele, "Oxygen Ion Conductors and Their Technological Applications," *Mater. Sci. Eng. B*, **13**, 79-87 (1992).

[12]J. Mizusaki, Y. MiMa, S. Yamauchi, K. Fueki, and H. Tagawa, "Nonstoichiometry of the Perovskite-Type Oxides $La_{1-x}Sr_xCoO_{3-\delta}$," *J. Solid State Chem.*, **80**, 102-11 (1989).

[13]J. Mizusaki, H. Tagawa, K. Naraya, and T. Sasamoto, "Nonstoichiometry and Thermochemical Stability of the Perovskite-Type $La_{1-x}Sr_xMnO_{3-\delta}$," *Solid State Ionics*, **49**, 111-18 (1991).

[14]M.H.R. Lankhorst and J. E. ten Elshof, "Thermodynamic Quantities and Defect Structure of $La_{0.6}Sr_{0.4}Co_{1-y}Fe_yO_{3-\delta}$ ($y = 0$-0.6) from High-Temperature Coulometric Titration Experiments," *J. Solid State Chem.*, **130**, 302-10 (1997).

[15]G.Ch. Kostogloudis, P. Fertis, and Ch. Ftikos, "Electronic Conductivity in the $Pr_{1-x}Sr_xCo_{1-y}Mn_yO_{3-\delta}$ System," *Solid State Ionics*, **118**, 241-49 (1999).

[16]K. Huang, H.Y. Lee, and J.B. Goodenough, "Sr- and Ni-Doped $LaCoO_3$ and $LaFeO_3$ Perovskites : New Cathode Materials for Solid Oxide Fuel Cells," *J. Electrochem. Soc.*, **145**, 3220-27 (1998).

[17]C.N.R. Rao, O.M. Parkash, and P. Ganguly, "Electronic and Magnetic Properties of $LaNi_{1-x}Co_xO_3$, $LaCo_{1-x}Fe_xO_3$, and $LaNi_{1-x}Fe_xO_3$," *J. Solid State Chem.*, **15**, 186-92 (1975).

[18]H. Hayashi, M. Kanoh, C. J. Quan, H. Inaba, S. Wang, M. Dokiya, and H. Tagawa, "Thermal Expansion of Gd-Doped Ceria and Reduced Ceria," *Solid State Ionics*, **132**, 227-33(2000).

ANODE SUPPORTED SOLID OXIDE FUEL CELLS WITH IMPROVED CATHODE/ELECTROLYTE INTERFACE

D. Montinaro[a], S. Modena[c], S. Ceschini[c], M. Bertoldi[b], T. Zandonella[b], A. Tomasi[c], V. M. Sglavo[a]

[a]*DIMTI, Università degli Studi di Trento, via Mesiano 77 38050 Trento, Italy*
[b]*Eurocoating SpA, via Al Dos de la Roda 60, Pergine Valsugana (TN), Italy*
[c]*ITC-Irst, via Sommarive 18, 38050 Povo (TN), Italy*

ABSTRACT

In the present work SOFCs with Ni based cermet anode and 8YSZ electrolyte were prepared by tape casting and co-sintering of laminated green layers at 1400°C. In addition, in order to increase the extent of TPBs, a porous 10YSZ interlayer was inserted at the electrolyte/cathode interface. An LSM20 gel precursor was prepared by polymerizable complex route and a composite cathode was then obtained by impregnation of the porous electrolyte with the LSM20 gel precursor and calcination. A current collector was screen printed on the cathode layer and sintered at 1150°C. In spite of the thermical expansion coefficient mismatch and the different sintering behaviour, micro-structural investigation showed a perfect adhesion between co-sintered electrodes and electrolyte. Open Circuit Voltage (OCV) measurements were performed in the temperature range 750-900°C and a perfect gas-tightness was confirmed by an high OCV value. A maximum power density of 230 $mWcm^{-2}$ was obtained working at 900°C under static air condition and using dry H_2 pure flux.

INTRODUCTION

Planar anode supported cells based on thin stabilised zirconia electrolyte are the most promising solid oxide fuel cells due to their well-proven long term stability [1] and good thermal expansion matching with low-cost ferritic stainless steel used as interconnectors for intermediate temperature applications [1-4].

$(La,Sr)MnO_3$ (LSM) perovskites are well known SOFCs cathode materials because of their electrocatalytic activity for oxygen reduction and their compatibility for Yttria-Stabilized Zirconia (YSZ)-based electrolytes [1-4]. Anode supported solid oxide fuel cells with LSM cathodes and 5-10 μm YSZ thick electrolytes still achieve good performance in the temperature range 750-850°C the cast overpotential being the most limiting factor on cell performance.

LSM polarization behaviour under defined working conditions is known to be related to oxide stoichiometry [5-9], electrode morphology [5-9] and extent of LSM/electrolyte interface area [10].

Due to the well known predominant electronic conductivity of LSM, mixed conductive ionic-electronic cathodes are typically made by electronic conductive perovskite oxide and ionic conductive electrolyte oxide [1-9]. Therefore, conventional composite cathodes are realized by using a mixture of LSM20 and 8YSZ particles. Such solution allows to increase the three phase boundaries (TPBs) at the electrolyte/cathode interface and decrease the thermal expansion coefficient mismatch between YSZ and LSM. Electrolyte and cathode particles are generally mixed in 30/70 or 50/50 volume ratios and coated on the electrolyte surface [9]. Composite cathodes are also fabricated by impregnating a sintered porous electrolyte interlayer with a perovskite precursor solution [11].

Water based tape casting is an attractive technology to produce SOFC materials it being an easy process, environmental friendly and a health-safe method. However, due to thermal expansion coefficient mismatch between electrodes and electrolyte, fuel-cells fabrication typically requires several ceramic depositions and sintering steps. Tape casting technology has been widely used to prepare SOFCs supporting electrodes [12-17] but this method was not extensively employed in the fabrication of electrolytes with thickness below 10 μm due to difficulties involved in obtaining a gas non-permeable thin ceramic layer.

In the present work SOFCs with Ni based cermet anode and 8YSZ electrolyte were prepared by tape casting and co-sintering of laminated green layers. In addition, in order to increase the extent of TPBs, a porous 10YSZ interlayer was inserted at the electrolyte/cathode interface. A composite cathode was then obtained by impregnation of the porous electrolyte with a LSM20 gel precursor current collector was screen printed on the cathode layer.

EXPERIMENTAL

Water based ceramic slurries containing dispersant (Darvan 821A, R.T. Vanderbilt Inc., USA) and ceramic powders (NiO/8YSZ and 8YSZ, NexTech Materials, USA) were milled 12 h by zirconia balls (Inframat Advanced Materials), filtrated to 40 μm and de-gassed by vacuum application. Then binders (B-1000 and B-1014, Duramax, Rohm and Haas, France) were added and the mixtures were mixed for 2 h by mechanical stirrer. Slurries were casted by a laboratory scale tape casting equipment. Graphite (Alfa Aesar, 7-10 micron flakes) was added to the anodic mixture in order to increase porosity. Anodic tapes without graphite was also prepared to provide a large distribution of three phase boundaries (TPBs) at the anode/electrolyte interface, to aid electrolyte sintering and to enhance mechanical strength.

In addition, in order to obtain a very porous zirconia layer close to the cathodic interface of the electrolyte, a tape made by fibrous 10YSZ particles (ZYP-55, Zircar Products Inc., USA) and graphite was prepared. Slurries compositions and casting parameters are reported in Tables I and II, respectively. Casted tapes were dried at T=25°C under RH 70-90% conditions.

Tabella I Ceramic slurries vol% composition (binders and dispersant ratios are based on dry polymers).

	A-p	A-d	E-d	E-i
Oxide powder	20	21	25	11
Water	57	58	67	69
Graphite	2	0	0	8
Darvan 821A	3	3	1	1
Duramax B 1000	13	13	3	8
Duramax B 1014	5	5	1	3

Table II. Green tapes casting parameters.

	Casting speed (m/min)	rear blade height (μm)	front blade heigh (μm)
A-p	0.5	500	500
A-d	1	150	150
E-d	2	50	25
EC-i	2	50	25

Anode, electrolyte and porous zirconia layers were stacked together, laminated under 12 MPa pressure (Material Test System 810) at 80°C for 30 min. Green laminates were co-fired 6 h at 1400°C and the as sintered half cells were "warm flattened" (1400°C) by the application of a dead-load which develops a pressure of about 6 MPa. Green layers sequence before lamination was: porous anode (labeled A-p; 4 layers, 230 μm each), dense anode (labelled as A-p; single layer, 80 μm), dense electrolyte (labeled E-d; single layer, 12 μm), and porous YSZ electrolyte-cathode interlayer (labeled EC-I; 1-2 layers, 12 μm each).

Electrolyte gas tightness was tested by a simple colorimetric method. A 30 x 30 mm half cell disk was sealed by a double o-rings system on the neck of an Herlenmeyer flask containing 30% ammonia solution, few drops of phenolphthalein were poured on the electrolyte side and vacuum was applied. Interconnected pores are revealed by phenolphthalein switching to pink due to ammonia gas permeation through the electrolyte.

A-site non-stoichiometric LSM20 gel precursor was prepared by the well known polymerizable complex route [18-23] using a metal nitrates to citric acid (CA) and to ethylene glycol (EG) molar ratio: 1:4:2. La (Treibacher, 99.99%), Mn (Alfa Aesar, 99.98%) and Sr (Alfa Aesar, 98%) nitrates were dissolved in a 1.8M CA solution in a two-neck round-bottom flask equipped with Allhin condenser. After stirring for 2 h at 80°C EG was added to the mixture and the solution was stirred for 4 h at 120 °C.

EC-i was impregnated by LSM20 gel solution and vacuum (11 mmHg) was applied to remove the solvent and to promote gel permeation through the whole porous structure.

On the basis of DTA measurements (Fig. 1) impregnated cells were heated for 1 h at 450°C to complete the organic compounds pirolysis and calcined for 6 hrs at 900°C.

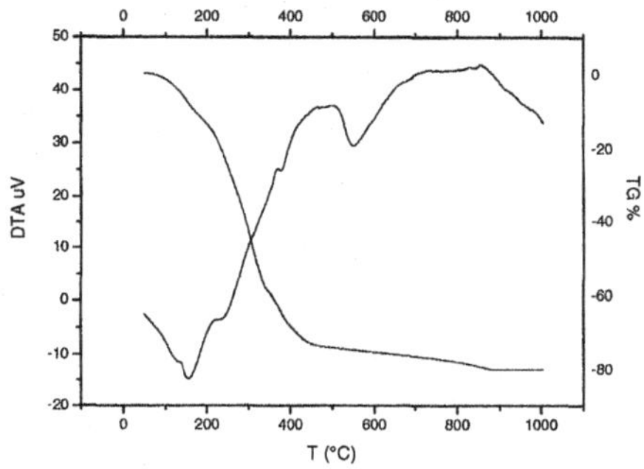

Figure. 1: TG-DTA analysis on LSM precursor powder.

A 50 μm current collector was screen printed onto the cathode and sintered for 2 h at 1150°C. The screen printing paste was produced by dispersing 60% LSM20 powders (Nextech Materials Ltd.,USA) in a mixture of therpineol and ethylcellulose.

A part of the LSM gel solutions was dehydrated at 150°C, placed in a 8YSZ crucible and submitted to the same heat treatment. X-rays diffraction analysis was performed on the resulting powders.

Micro-structural analysis were performed by JEOL JSM 5500 Scanning Electron Microscope (SEM).

The performances of fuel cells were measured in a testing fixture consisting of an alumina tubular furnace in which a chemical reactor with the button cell loaded at the inner end is inserted; another small alumina tube is used to carry the reactive gas. Gas tight stainless steel faceplates are used at the outer end of alumina tubes; the faceplates, such as drilled gas inlet and outlet, voltage and current connectors were sealed using silicone O-rings.

A k-type thermocouple was also used to obtain accurate fuel cell temperature.

The samples as square cells are loaded in vertical position at the inner end of the chemical reactor. The cell was mounted and sealed to the alumina tube by using alumina rings. The anode side of the cells was bonded to the rings and the edges of the cell were sealed using ceramic pastes.

Platinum meshes were used on both electrodes to collect the electrons. A few drops of Pt ink were also used to ensure the electrical contact between the mesh and the electrode surfaces.

The cathode side of the cell was exposed to static air, while the anode side was exposed to the flowing reactive gas.

Separate conductive wires were used to carry the current and voltage of the cell. The contacts between the wires and platinum meshes, both on anode and cathode sides, were ensured using Pt ink.

The I/V behavior was studied by a Fuel Cell Testing Module, consisting of an ammeter and voltmeter with different adjustable resistances and of a power supply.

The cell was heated up to 800°C in inert atmosphere in order to remove oxygen molecules from the reactive chamber and to cure the ceramic-based sealants and Pt-ink paste. At this time the resistance between the voltage and current leads of each electrodes were measured to ensure each wire was still connected to the Pt mesh.

Nerst potential was then measured. This is very sensitive to changes in partial pressures; therefore, its measurements is an efficient method to reveal the presence of small cracks or flaws.

After the completion of such first test carried out to ensure the absence of leaks, a reactive gas (90% inert gas – 10% H_2) was introduced to the anode to perform the reduction. The NiO/YSZ anodes were reduced into Ni/YSZ cermet at temperatures in the range 700-800°C for several hours before open circuit voltages were measured. After reduction, electrochemical measurements were carried out from 750°C to 900°C in 50°C intervals using dry H_2 pure flux.

RESULTS AND DISCUSSION

The half cells shrinkage after sintering treatment was observed to be in the order of 23%.

Microstructure investigation was performed before and after electrochemical measurements and perfect adhesion was observed at each interface in both reduced and non reduced cells.

Sintered electrolyte layer was in the range of 5-10 μm and no interconnected porosity was observed. A perfect gas-tightness was also confirmed by phenolphtalein test.

XRD measurements performed on LSM powders calcined at 900°C and 1150°C respectively reveal only pure perovskite phase (Fig. 2).

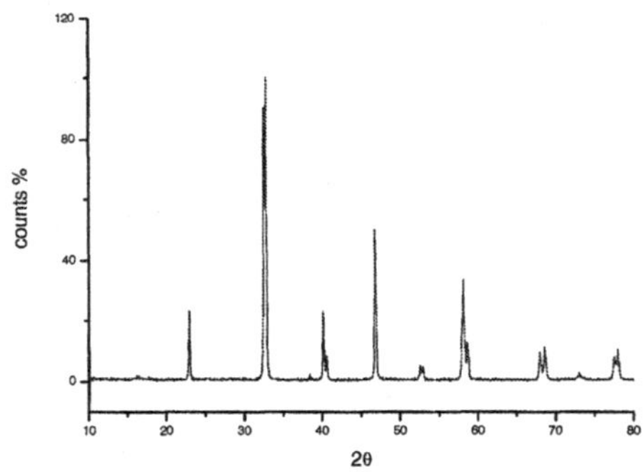

Figure. 2 . XRD pattern of calcined LSM20 obtained by pirolysis of the citrate gel precursor at 450°C and calcination at 900°C for 6 h.

In order to obtain a very dense electrolyte layer after sintering process, 8YSZ slurries was prepared with a very small amount of acrylic binders while in anode and EC-i tapes the B1000 and B1014 ratio was optimized to allow a perfect green layers adhesion under warm pressing conditions.

Due to graphite deformation during both tape casting process and uni-axial warm-pressing big pores appear in the anode side slightly oriented along the fuel cell plane.

Despite to the fibrous shape of 10YSZ powders used in the slurry formulation, EC-i shows a very interconnected structure and no particles orientation was observed (Fig. 3-5). The high graphite content may increase the layer compaction and particle packing under warm-pressing and this is a possible explanation of the observed high thickness reduction of the sintered 10YSZ layer upon sintering (Fig. 3). Hence several green layers of 10YSZ were needed to prepare the half cells with different 10YSZ interlayer thickness (Fig. 3, 4).

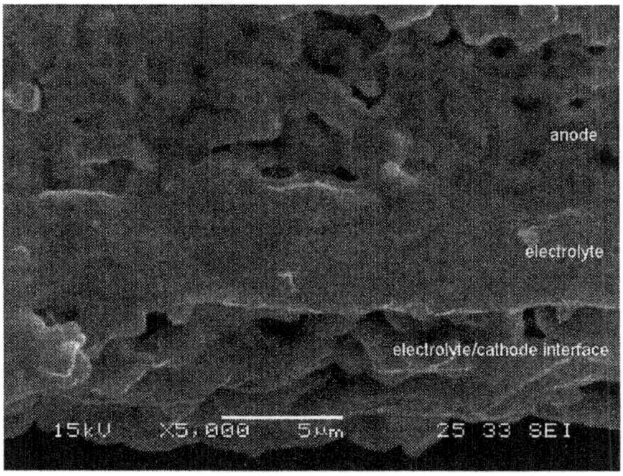

Figure. 3Microstructure of the enhanced electrolyte/cathode interface area.

Figure 4. SEM micrograph of the elecrolyte/cathode interface fracture surface after gel calcination to 900°C. Fine particles are LSM20.

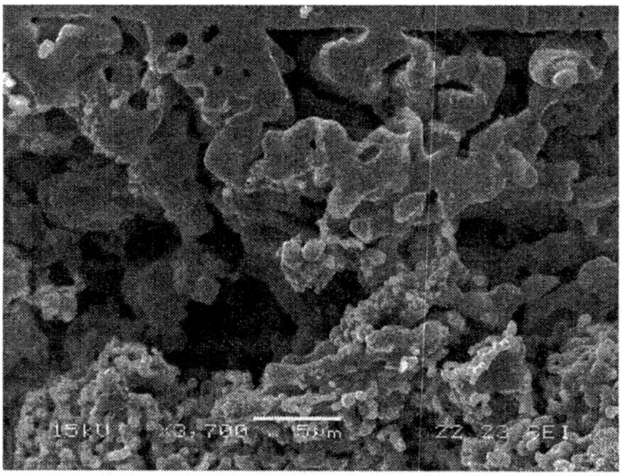

Figure 5. SEM micrograph of the impregnated electrolyte/cathode interface after current collector deposition and sintering at 1150°C.

Due to the porous nature of the increased interfacial area, impregnation was considered the more suitable method to obtain a composite electronic-ionic conductive cathode.

SEM investigations performed on half-cells impregnated with the LSM precursor and calcined at 900°C show a perfect filling of the EC-i and an interconnected structure of LSM20 fine particles.

However, after current collector deposition and sintering at 1150°C, LSM particles produced from citrate gel combustion show a large shrinkage effect and a poor interconnected structure was obtained (Fig. 3). Therefore, fuel-cells were again impregnated with LSM20 precursor and calcined at 900°C to fill the interlayer with fine and interconnected LSM particles (Fig. 4).

In order to prevent Ni re-oxidation to NiO under close circuit condition, measurements were limited at a cut-off voltage of 450 mV and were performed under static air condition. Results of electrochemical measurements are summarized in Table III. The high Open Circuit Voltage (OCV) values observed in the temperature range between 750 and 900°C confirm the perfect gas-tightness of the thin dense electrolyte layer prepared by tape casting and co-sintered with the supporting anode and the porous EC-i layer.

Voltage/Current-Density (V/I) curves for different testing temperatures are reported in Fig.6. Due to the decrease of ohmic resistance on increasing working temperature a considerable reduction of curves slope was observed. No appreciable differences were observed in slopes of V/I curves in the range 750-800°C. Power density curves (Fig. 7) shows a maximum value of 230 $mWcm^{-2}$ for a cell working at 900°C on static air and pure hydrogen condition and at 210 mV. Such result is appreciable but not still comparable with results obtained by conventional composite cathodes[1-9].

Table III: Results of the electrochemical measurements performed on fuel cells.

T (°C)	OCV (mV)	ASR[†] ($m\Omega\ cm^{-2}$)	P[‡] ($mW\ cm^{-2}$)
750	1091	5	20
800	1085	5	120
850	1080	4	159
900	1064	3	230

[†] Area Specific Resistance measured at open circuit voltage (OCV)= 700 mV.

[‡] Maximum power density at a 450 mV cut-off voltage.

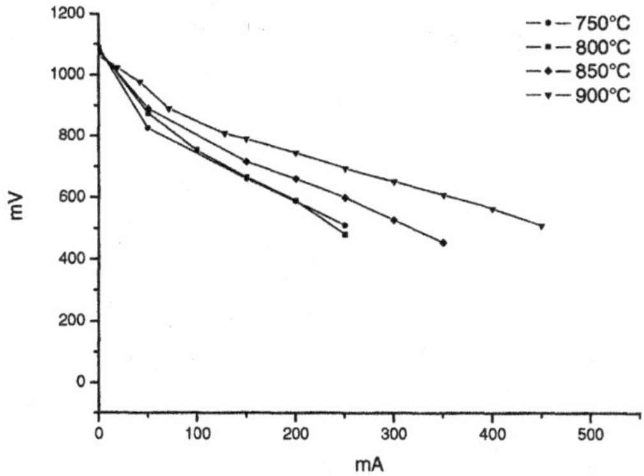

Figure 6. Voltage/Current-Density measurements performed under static air condition and pure H_2 flux.

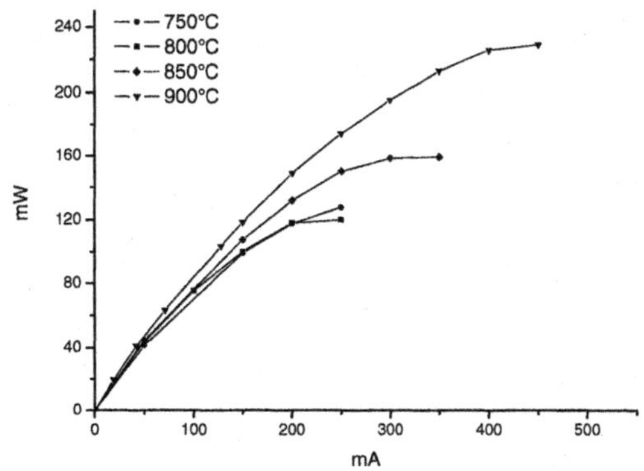

Figure. 7. Power-density curves for a cut-off voltage of 450 mV.

The limited performances were probably due to a non-adequate LSM structure inside the porous electrolyte layer but the feasibility of one-step producing half cells with a 5 μ thin dense electrolyte and with an enhanced electrolyte/cathode interface area encourages to optimize the impregnation process in order to decrease cathodic polarization losses.

CONCLUSIONS

NiO/YSZ based anode supported fuel cells with a 5-6 μ thick gas-thigh 8YSZ electrolyte were successfully prepared by water based tape casting and co-sintering. In addition cathode/electrolyte interface area was increased by preparing a layer with fibrous 10YSZ powders. In spite of the thermal expansion coefficient mismatch and the different sintering behaviour, micro-structural investigation showed a perfect adhesion between co-sintered electrodes and electrolyte. Phenolphtalein test and Open Circuit Voltage measurements confirmed the perfect gas-thigh behaviour of sintered half cells. A composite cathode was then obtained by impregnation of the EC-i matrix with a LSM gel precursor obtained by polymerizable complex route and calcination at 900°C. A maximum power density of 230 mWcm^{-2} was obtained working at 900°C under static air condition and using dry H_2 pure flux. The feasibility to prepare SOFC with a very thin electrolyte by a technology based on water slurries and easily scaling up to large area scale is an attractive feature and encourages to devote further studies on the cell structure optimization in order to achieve better performances.

REFERENCES

1. E. Ivers-Tiffée, A. Weber and D. Herbstritt, "Materials and technologies for SOFC-components", *J. European. Ceram. Soc.*, **21**, 1805-1811 (2001)
2. W. J. Quadakkers, H. Greiner, M. Hansel, A. Pattanaik, A.S. Khanna and W. Mallener, "Compatibility of Perovskite Contact Layers Between Cathode and Metallic Interconnectror plates of SOFCs", *Solid State Ionics*, **91**, 55-67 (1996)
3. T. Brylewki, M. Nanko, T.Maruyama, K. Przybylski "Application of Fe–16Cr Ferritic Alloy to Interconnector for a Solid Oxide Fuel Cell", *Solid State Ionics*, **143**, 131-150 (2001)
4. B. C. H. Steele, "Materials for IT-SOFC stacks 35 years R&D: the inevitability of gradualness?" *Solid State Ionics*, **134**, 3-20 (2000).
5. S. P. Jiang, J. G. Love, J. P. Zhang, M. Hoang, Y. Ramprakash, A. E. Hughes and S. P. S. Badwal, "The Electrochemical Performance of LSM/Zirconia-Yttria interface as a function of A-Site Non-Stoichiometry and Cathodic Current Treatment", *Solid State Ionics*, **121** 1-10 (1999)
6. S. P. Jiang, J. P. Zhang, Y. Ramprakash, D. Milosevic adn K. Wilshier, "An investigation of Shelf-Life of Strontium Doped $LaMnO_3$ Materials", *J. Mater. Res.*, **35**, 2735-2741 (2000).
7. S. P. Jiang, "Issue on Development of $(La,Sr)MnO_3$ Cathode for Solid Oxide Fuel Cells", J. Power Sources, **124**, 390-402 (2003)
8. S. P. Jiang, J. P. Zhang, K. Foger, "Chemical Interactions Between 3 mol% Yttria-Zirconia and Sr-Doped Lanthanum Manganite", *J. European Ceram. Soc.*, **23**, 1865-1863 (2003)
9. K. Bartthel and S. Rambert, "Processing and Electrochemical Behaviour of LSM-YSZ Composite Cathodes by Vacuum Plasma Spraying", 3rd European Solid Oxide Fuel Cell Forum, Nantes, F, 11-18 (1998).
10. V.Brichzin, J. Fleig, H.-U. Habermeier and J. Majer "Geometry Dependence of Cathode Polarization in Solid Oxide Fuel Cells Investigated by Defined Sr-Doped $LaMnO_3$ Microelectrodes" *Electrochem. Solid State Lett.*, **3**, 403-406 (2000).
11. H. He, Y. Huang, J. Regal, M. Boaro, J. M. Vohs and R. J. Gorte, "Low-Temperature Fabrication of Oxide Composites for Solid-Oxide Fuel Cells", **87**[3], 331-336 (2004).
12. Park, R. J. Gorte and J. M. Vohs, "Tape Cast Solid Oxide Fuel Cells for the Direct Oxidation of Hydrocarbons", *J. Electrochem. Soc.*, **148**[5], A443-A447 (2001)
13. H. Kim, C. da Rosa, M. Boaro, J. M. Vohs and R. J. Gorte, "Fabrication of Highly Porous Yttria-Stabilized Zirconia by Acid Leaching from a Nickel-Yttria-Stabilized Zirconia Cermet", *J. Am. Ceram, Soc.*, **85** [6] 1473-1476 (2002).
14. M. Boaro, J. M. Vohs and R. J. Gorte, "Synthesis of Highly Porous Yttria-Stabilized Zirconia by Tape-Casting Methods, **86** [3] 395-400 (2003).
15. S. Primdahl, B. F. Sørensen and M. Mogensen, "Effect of Nichel Oxide/Yttria-Stabilized Zirconia Anode Precursor Sintering Temperature on the Properties of Solid Oxide Fuel Cells", **83**[3] 489-94 (2000).
16. AC. Müller, D. Herbstritt and E. Ivers-Tiffée,"Development of Multilayer Anode for Solid Oxide Fuel Cells", *Solid State Ionics*, **152-153**, 537-542 (2002).
17. D. Simwonis, H Thülen, F.J. Dias, A. Naoumidisa and D. Stöver, "Properties of Ni/YSZ Porous Substrates Prepared by Tape Casting and Coat-Mix Process", J. Materials Processing and Technology, **92-93**, 107-111 (1999)

18. M. Pechini, "Method of Preparing Lead and Alkaline Earth Titanates and Niobates and Coating Method Using the Same to Form a Capacitor," U.S. Pat. No. 3 330 697, July 11, 1967.
19. F. Licci, G. Turilli, P. Ferro and A. Ciccarone, "Low-Temperature Synthesis and Properties of $LaMnO_{3\pm d}$ and $La_{0.67}Sr_{0.33}MnO_{3\pm d}$ (R = Ca, Sr, Ba) from citrate Precursors", *J. Am. Ceram, Soc.*, **86** [3], 413-419 (2003).
20. M. Popa and M Kakihana, "Synthesis of Lanthanum Cobaltite (LaCoO3) by the Polymerizable Complex Route", *Solid State Ionics*, **151**, 251-257 (2002).
21. M. Popa L. Van Hong and M Kakihana, "Nanopowders of LaMeO3 perovskites obtained by a Solution-Based Ceramic Processing Technique", *Physica* B, **327**, 233-236 (2003).
22. Z. Peng and M. Liu, "Preparation of Dense Platinum-Yttria Stabilized Zirconia and Yttria Stabilized Zirconia Films on Porous $La_{0.9}Sr_{0.1}MnO_3$", *J. Am. Ceram, Soc.*, **84**, [2] 283-28 (2001).
23. E. Krupicka, A. Reller and A. Weidenkaff, "Morphology of Nanoscaled $LaMO_3$-Particles (M=Mn, Fe, Co, Ni) Derived by Citrate Precursor in Aqueous and Alcoholic Solvents", *Crystal Engineering*, **5**, 195-202 (2002).

LONG-TERM EFFECTS IN Ag-CuO BRAZES UNDER DUAL REDUCING/OXIDIZING GAS CONDITIONS

K. S. Weil, J. Y. Kim, and J. S. Hardy
Pacific Northwest National Laboratory
P.O. Box 999
Richland, WA 99352

ABSTRACT

A new brazing technique is currently being developed specifically for sealing solid state electrochemical devices such as planar solid oxide fuel cells (pSOFCs). Referred to as air brazing, this metal-based joining technique is unique in that it can be conducted directly in air without the use of inert gases or fluxing agents to prevent/remove surface scale during the joining process. To date, several developmental silver-based filler metal compositions have shown good joint strength, excellent stability under high-temperature oxidizing conditions, and superior thermal cycling performance in comparison with glass-based seals. However, there are concerns about exposing silver alloys to a high-temperature, hydrogen-rich environment, particularly with respect to potential embrittlement under these conditions. This paper discusses initial results obtained from a series of experiments designed to characterize the effect of simultaneous hydrogen/air exposure on silver-based air brazes at high temperature.

INTRODUCTION

Because pSOFCs function under an oxygen ion gradient that develops across the electrolyte, hermiticity across this membrane is paramount. In a planar design, this means that the YSZ layer must be dense, must not contain interconnected porosity, and must be connected to the rest of the device with a high temperature, gas-tight seal. One of the fundamental challenges in fabricating pSOFCs is how to effectively seal the thin electrochemically active YSZ membrane against the metallic body of the device to create a hermetic, rugged and stable stack. Typical conditions under which these devices are expected to operate and to which the accompanying YSZ-to-metal seal will be exposed include: (1) an average operating temperature of 750°C; (2) continuous exposure to an oxidizing atmosphere on the cathode side and a wet reducing gas on the anode side; and (3) an anticipated device lifetime of 10,000+ hours.

While a number of sealing techniques[1-4] are under consideration for pSOFC stack design, including glass bonding,[1,2] compressive sealing,[3] and reactive joining,[4] one of the most reliable and best understood methods of joining dissimilar materials is brazing.[5] Recently a new method of ceramic-to-metal brazing has been developed specifically for use in fabricating high-temperature solid-state devices such as pSOFCs and oxygen and hydrogen concentrators.[6] Referred to as air brazing, the technique differs from traditional ceramic brazing in two important ways: (1) it utilizes a liquid-phase oxide/noble metal melt as the basis for joining, and thereby exhibits inherent high-temperature oxidation resistance, and (2) the process can be conducted directly in air without the use of fluxes and/or inert cover gases. In fact, the strength of the bond formed during air brazing relies on the formation of a thin, adherent oxide scale on the metal substrate. One system that has shown particular promise in the development of air brazing is Ag-CuO. Prior work has shown that the addition of 1.4 to 8 mol% CuO to silver results in excellent wetting on a variety of oxide and oxidized metal surfaces, while producing filler metal and interfacial microstructures that lead to good joint strength, excellent oxidation resistance, and superior thermal aging and thermal cycling properties, particularly in comparison with barium aluminosilicate-based glass sealed joints.[6,7]

However, there are several concerns with using silver-based alloys in pSOFC applications, including: (1) silver volatility, (2) degradation under dual atmosphere exposure, and (3) electrical shorting through the seal. With respect to dual atmosphere exposure, the issue is silver embrittlement when the metal is simultaneously exposed to oxygen (air) and hydrogen. Both species are relatively soluble in silver and display rapid rates of diffusion within the interstitial void space of pure metallic silver. Klueh and Mullins were the first to observe that under certain conditions water vapor bubbles will form both within the grains and along the grain boundaries of silver, ostensibly due to reaction between the two diffused species.[8] In the present study, a series of dual atmosphere exposure experiments were conducted to examine this effect in Ag-CuO braze alloys and understand how degradation might take place.

EXPERIMENTAL

Alumina cap and pipe specimens of the type shown in Figure 1(a) were employed in exposure testing of the Ag-CuO seal. Although in the SOFC stack it envisioned that air braze seal will be between YSZ and ferritic stainless steel, alumina serves as a useful inert substrate in which the effects of dual atmosphere exposure can be investigated. The dual atmosphere exposure condition was established by heating the braze sealed end of the pipe in air while circulating wet H_2 gas within the pipe. As shown in Figure 1(b), the cold end of the pipe that extends outside of the muffle furnace was sealed using Swagelok® compression fittings to ensure that the hydrogen could be supplied to the in-board side of the seal via a central tube and exhausted via a port without any external leaks. High purity (99.5% Al_2O_3, ~99% dense; Coors Co.) alumina pipes measuring 1.25 inches in diameter x 18 inches long with a 1/8" wall thickness were used in the study. The alumina caps employed were also high purity (99.7% Al_2O_3, 98% dense; Alfa Aesar) and measured approximately 1.25 inches in diameter by 1/8" thick. The faying surfaces of both the pipes and caps were polished to a 10μm finish using successively finer grit diamond paste, cleaned with acetone, and rinsed with propanol, air dried, and finally heated in static air to 600°C for four hours to burn off any residual organic contamination.

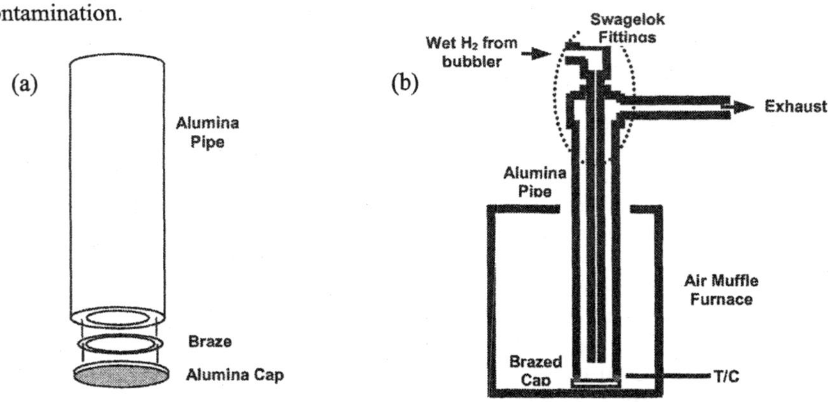

Figure 1 Schematics of (a) the braze sealed alumina pipe specimen and (b) the dual atmosphere exposure test configuration.

The air braze filler metal composition employed (4mol% CuO in silver; Ag4CuO) was prepared by electroplating a 300μm thick silver foil in a 1M aqueous $CuSO_4$ + 1M H_2SO_4 solution at 0.1A (limited to 0.3V) and 70°C. The period of deposition was based on the final target composition of the filler material, taking into account that the plated copper will oxidize in-situ during the brazing process. Once the foils were plated, they were homogenized at 700°C for 10hrs in Ar/2.75% H_2 to form a silver-copper alloy. During heat treatment, the foils wrinkle slightly, so they were rolled to a final thickness of 120μm to ensure flatness, then cut into appropriately sized rings for brazing. Joining was conducted by placing a braze foil ring on the faying surface of an alumina pipe, stacking an alumina cap on top of the foil, and adding a 50g dead-load weight on the cap to ensure good contact during brazing. The assembly was heated in air at 5°C/min to 1100°C and held at temperature for 30min before furnace cooling to room temperature.

Exposure testing was conducted by placing the brazed end of the pipe into an air muffle furnace, connecting the compression fittings at the cold end, and flowing UHP H_2 bubbled through water into the pipe via a central inlet tube. The brazed portion of the specimen was heated in static air at 3°C/min to 750°C and held at this temperature for the duration of the test. Once high-temperature exposure was complete, the specimen was cut ~5mm from the braze bond line to remove the cap, the joining region, and the adjacent pipe material. The resulting piece was sectioned for microstructural analysis. Examination was performed using a scanning electron microscope (SEM, JEOL JSM-5900LV). The SEM is equipped with an Oxford energy dispersive X-ray analysis (EDX) system, which employs a windowless detector for quantitative detection of both light and heavy elements. In order to avoid electrical charging on the samples, they were carbon coated and grounded. Elemental profiles were determined across the joint interfaces in the line-scan mode.

RESULTS AND DISCUSSION

For the sake of comparison, an air brazed alumina/alumina joint prepared using a braze composition of 4mol% CuO in Ag is shown in the as-joined condition in Figure 2. In this particular example, the sample exhibits no porosity at the interface. In fact the filler material infiltrates into the submicron roughness on the alumina faying surface, indicating good bonding between the braze and substrate. The copper oxide found in the joining region exists as precipitates along the alumina-braze interface and as equiaxed particulate in the bulk filler metal region, which otherwise consists of nearly pure silver. A small amount copper oxide has also reacted with the alumina substrate forming copper aluminate, $CuAlO_2$. The CuO/Al_2O_3 phase diagram indicates that this reaction is thermodynamically possible at temperatures in excess of ~1100°C.[9]

Shown in Figure 3 are the microstructural results obtained on a joining specimen that was exposed to the 750°C, dual flowing H_2/static air environment for a period of 260hrs. This particular seal measures 3.2mm in width by ~95μm in thickness. As indicated in the figure, some porosity has formed within the joint along the hydrogen exposure side during testing and extends ~450μm into the joint. The pores in this region are uniformly spaced and measure approximately 2 – 8μm in diameter. A fairly distinct line of demarcation is observed within the filler metal, beyond which towards the air exposure side of the joint virtually no porosity is evident. Also apparent on either side of this line is the presence of (or lack thereof) CuO precipitates in the bulk of the filler metal and along the interfaces with the alumina. On the "hydrogen" side of this

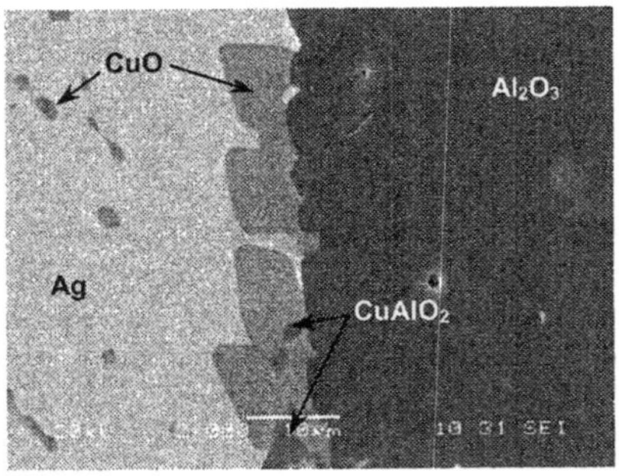

Figure 2 A typical air brazed alumina-alumina joint prepared using a 4mol% CuO in Ag composition.

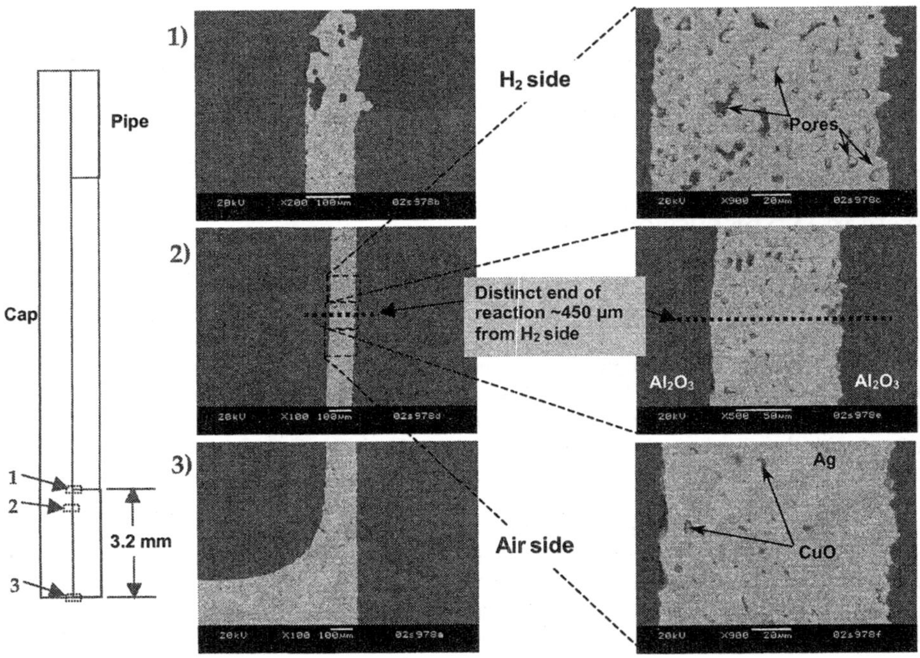

Figure 3 Cross-sectional microstructure of a pipe and cap specimen that was exposure tested under external static air and internal wet flowing hydrogen for 260hrs at 750°C.

line, little CuO is found, whereas on the "air" side the CuO precipitates appear as before in the as-joined sample of Figure 2. In general, the joint is sound and the porosity observed is non-percolating. Post-exposure leak testing using a slightly positive pressure of He internally within the sealed pipe and a He leak tester external to the brazed joint indicated no loss in hermeticity across the seal due to dual atmosphere exposure. The seal displayed no apparent signs of bloating or significant geometric change due to internal pore formation.

Microstructural results obtained on a second air brazed pipe and cap specimen are shown in Figure 4. This specimen was also exposure tested at 750°C under a dual flowing H_2/static air environment, in this case for a period of 550hrs. The brazed seal measures 3.1mm in width by ~25μm in thickness. Again, uniform pore formation takes place in that portion of the filler metal lying adjacent to the hydrogen exposed free surface. The pores measure approximately 3 - 10μm in diameter. Virtually no CuO is found in this region of the joint, although some interfacial $CuAlO_2$ remains present. Given that the size of the aluminate precipitates is approximately the same as that observed in the as-joined specimen, it is initially concluded that this reaction product does not undergo reduction on the hydrogen side of the joint. This agrees with previous observations of $CuAlO_2$ in hydrogen exposed air braze joints.[10] The porosity extends approximately 650μm into the joint, beyond which the pores are generally absent. Again, the joint generally appears to be sound and the porosity remains non-percolating, with no signs of bloating and with post-exposure He leak testing indicating no loss in seal hermeticity.

In both samples, there was evidence of hydrogen diffusion into the filler metal region, causing localized reduction of the CuO. This process is accompanied by the formation of pores, or more accurately steam bubbles, via the following reaction:

$$CuO + 2H\,(\text{dissolved}) \rightarrow Cu + H_2O\,(v) \tag{1}$$

Because silver displays little solubility for water vapor, a small bubble containing the gas will form adjacent to the oxide precipitate as the reduction reaction proceeds. The copper metal produced in this reaction is free to dissolve and alloy into the neighboring matrix. Results from EDX analysis conducted on local regions of the filler metal lying adjacent to the pores indicate that this is the case. An example of the EDX findings is given in Figure 5. Although reduction occurs in part of the braze filler metal, there are no signs of de-wetting along interfaces with the alumina substrates as a result of this reaction. The filler metal remains intimately adhered to both sufaces as seen in Figures 3 and 4. Conversely, the $CuAlO_2$ phase found dispersed along the substrate interfaces within the joint does not appear to undergo reduction nor was pore formation near the air exposed surface of the brazed seal observed in these samples.

Noting that at 750°C the width of the reduction zone is ~450mm after 260hrs of exposure and extends to ~650mm after 550hrs, a simple diffusional calculation using $x \sim t^{1/2}$ (where x is the width of the porous zone and t is time of exposure) suggests that the two findings are in very good agreement. Although the results are quite preliminary and require additional testing and analysis, they suggest that if the mechanism of bubble formation remains constant the porosity would extend completely through the entire ~3mm length of the seal after ~11,500hrs of exposure. However, potentially there is an additional source of water vapor within the seal due to the reaction between diffused oxygen from the air side of the braze and hydrogen from the fuel side. Since the rate of hydrogen diffusion is much higher, it is expected that additional pore formation should take place near the air side of the braze where the two species would meet and

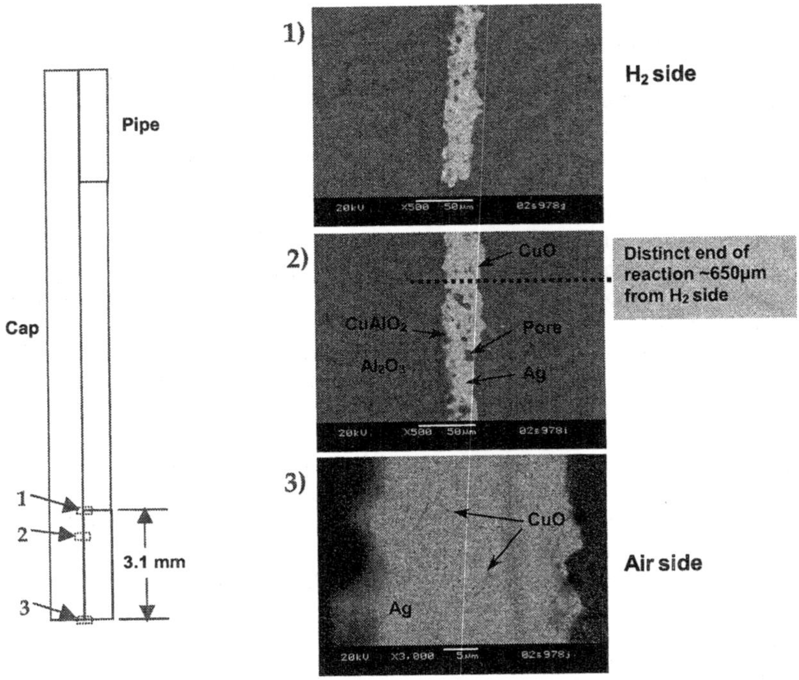

Figure 4 Cross-sectional microstructure of a pipe and cap specimen that was exposure tested under external static air and internal wet flowing hydrogen for 550hrs at 750°C.

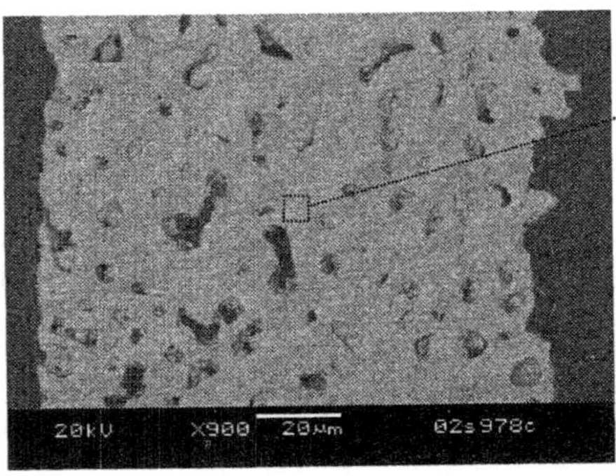

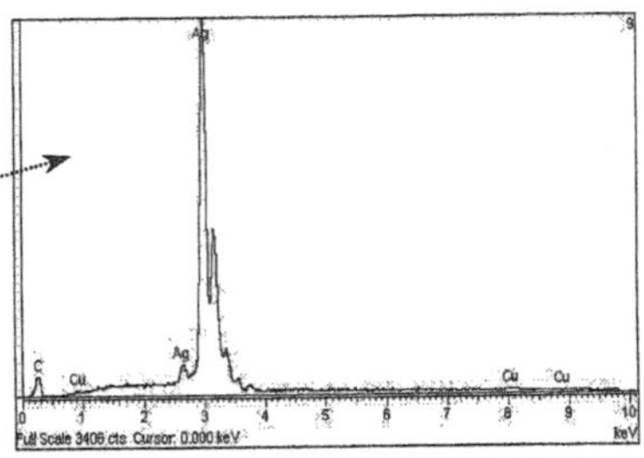

	Intensity	wt%	at%
Cu K	1.0431	2.28	3.76
Ag L	0.9841	97.74	96.24

Figure 5 EDX analysis of localized filler metal regions adjacent to pores found within the 550hr exposure specimen.

react, assuming that the hydrogen is not entirely consumed due to reduction of the CuO precipitates near the hydrogen exposure side of the joint. The fact that no pore formation is observed near the air side of the joint suggests that internal oxygen and hydrogen reaction does not occur in the air braze seal, at least out to 550hrs. Longer-term tests are currently being conducted to determine whether the mechanism of microstructural change within the braze remains the same at longer periods of dual atmosphere exposure.

CONCLUSIONS

One of the concerns of using Ag-CuO air brazed seals is currently being investigated. Prior studies on pure silver indicate that under a high-temperature dual atmosphere exposure, microstructural degradation takes place over time due to the simultaneous diffusion of hydrogen and oxygen into the metal and subsequent reaction, forming water vapor bubbles predominantly along grain boundaries. To test whether the same phenomenon occurs in Ag-CuO air brazed seals, brazed alumina pipe specimens were fabricated, connected to an internal source of flowing wet hydrogen, and placed into an air muffle furnace at 750°C for a specified period of time. After exposure, the specimens were He leak tested, then examined via SEM. The following effects were observed:

1. Reduction of the CuO precipitates commonly found in the braze filler metal occurs in that portion of the seal that lies adjacent to the free surface exposed to hydrogen. This leads to the formation of pores (i.e. water vapor bubbles) in this region of the seal.
2. The bubbles appear to be uniformly spaced and not interconnected.
3. A clear demarcation between the reduced portion of the braze and the unreduced portion is very evident. After 260hrs at 750°C, the reduction zone measures approximately 450μm in width, whereas after 550hrs at the same temperature it extends to 650μm.
4. The copper aluminate interfacial reaction product appears to be stable within the reduced side of the braze.
5. In both cases, the exposure tested brazed seal remains hermetic.

ACKNOWLEDGMENTS

The authors would like to thank Nat Saenz, Shelly Carlson, and Jim Coleman for their assistance. This work was supported by the U. S. Department of Energy as part of the SECA Program. The Pacific Northwest National Laboratory is operated by Battelle Memorial Institute for the United States Department of Energy (U.S. DOE) under Contract DE-AC06-76RLO 1830.

REFERENCES

[1] Z. G. Yang, G.G. Xia, K.D. Meinhardt, K.S. Weil, and J.W. Stevenson, *J. Mater. Eng. Perf.*, **13**, 327 (2004).
[2] Z.G. Yang, J.W. Stevenson, and K.D. Meinhardt, *Sol. St. Ionics*, **160**, 213 (2003).
[3] S.P. Simner and J.W. Stevenson, *J. Power Sources,* **102**, 310 (2001).
[4] A. K. Jadoon, *J. Mater. Sci.*, **39**, 593 (2004).
[5] C. W. Fox and G. M. Slaughter, *Welding J.*, **43**, 591 (1964).
[6] K. S. Weil, J. Y. Kim, and J. S. Hardy, *Electrochem. and Sol. St. Lett.*,**8**, A133 (2005).
[7] K. S. Weil, C. A. Coyle, J. T. Darsell, G. G. Xia, and J. S. Hardy, *J. Power Sources*, in press.
[8] D.A. R. L. Kleuh and W. W. Mullins, *Trans. Metall. Soc. of AIME*, 242, 237 (1968).
[9] S. K. Misra and A. C. D. Chaklader, *J. Am. Cer. Soc.*, **46**, 609 (1963).

[10]K. S. Weil and J. Y. Kim, in *Joining of Advanced and Specialty Materials VI*, ASM International, 2005.

SELF HEALING GLASS SEALS FOR SOLID OXIDE FUEL CELLS

Shailendra S Parihar and Raj N. Singh
Department of Chemical and Materials Engineering
University of Cincinnati
Cincinnati, OH-45221-0012

SUMMARY

Glass sealants can be used to hermetically seal different components of a solid oxide fuel cell stack. One of the limitations of glass sealant is its inherent brittleness, which can be overcome by utilizing the ability of a glass to self heal at higher temperatures. At these temperatures glass can flow and fill the cracks developed during thermal cycling. In this study, a glass of suitable composition, expansion and flow characteristics is used to seal yttria stabilized zirconia (YSZ) to a metal. The self healing behavior of these seals is demonstrated. Leak rate of these glass seals is measured at room temperature, higher temperatures and after thermal cycles.

INTRODUCTION

A fuel cell is an energy conversion device that produces electricity and heat by electrochemical combination of a fuel with an oxidant.[1] Among various types of fuel cells, solid oxide fuel cell (SOFC) offer advantage of highest efficiency, potentially long life, construction from readily available ceramic materials, production of high grade waste heat, possibility of internal reforming, and less problem with electrolyte management (unlike liquid electrolytes).[2]

There are different designs of SOFCs among which planar and tubular designs are more common. Planar design has advantage of mechanical robustness, high efficiency, high power density, and low cost.[3] In case of the planar type SOFCs, however, it is necessary to develop a suitable gastight seal for preventing the fuel gas and air from mixing during fuel cell operation.[4] In addition to gas tightness, seal should have mechanical stability (robustness under external forces), chemical stability in fuel cell environment, thermal cycling/thermal shock resistance, chemical inertness, acceptable bond strength, electrical insulation properties, and low cost.[5,6]

To fulfill these functional requirements sealing material should have very small thermal expansion mismatch with respect to other SOFC components, no chemical reactivity with joining components, stability under dual (oxidizing and reducing) environment, good wetting characteristics, proper viscosity at higher temperatures, and high electrical resistance.[7]

There are three main approaches for sealing SOFCs, rigid chemical seals (glass, glass-ceramics, high temperature cements, non-conducting brazes, non-oxide amorphous ceramics), sliding mechanical seals (mica, metal wire, ceramic fiber) and compliant wet seals. Some Advance approaches like self healing glass seals,[8,9] self healing ceramic composite seals, graded TEC seals, wet seals, layered composite seals,[10] and high temperature spring seals are also under development. Among all these approaches, rigid glass-based seal technology appears to be more mature and promising due to the low leakage, low cost, excellent thermal and environmental stability, processing flexibility, and successful short term experiences.[11,12]

Although, glass seals are lucrative option for sealing SOFC components yet they have some limitations including inherent brittleness of the glass. Due to the brittleness of the glass, cracks in the seal can be developed during thermal cycling or shock, which can cause leakage from the seal. One way to minimize the problem of glass brittleness is to use a glass that can show self healing at high temperatures.

In this study, a silica based glass is used as sealant. This glass demonstrated self healing behavior at fuel cell operating conditions, and thus enabled the seal to self heal the cracks developed in the glass because of the thermal cycling. An approach for selecting glass and its ability to self heal is demonstrated by leak testing.

EXPERIMENTAL

The sealing glass, designated as glass 4, is a SiO_2 based glass that has SiO_2, BaO, Na_2O, Al_2O_3, and other minor constituents. Glass tapes were made by tape casting. These tapes were laminated together to the final thickness. Glass laminates were sintered at 750^0C for 2 hours to make dense rectangular bars of 1 inch length for thermal expansion measurement and healing study. Sintered glass sample was used for density measurement and X-ray diffraction analysis.

The metal used in this study was a ferritic stainless steel known as Crofer 22 APU. Metal used in this study was 0.3 mm thick. For thermal expansion measurement metal sheet was rolled and pressed and 1 inch long sample was cut from it.

YSZ membranes were made using tapes fabricated from 8mol% yttria stabilized zirconia (YSZ) powder. After binder removal YSZ tapes were sintered at 1485^0C for 3 hours to make dense YSZ membrane. Dense samples of YSZ were used for making seals and expansion measurements.

Thermal expansion measurements on all the samples were done in an inert atmosphere at a heating rate of 3^0C/min using a differential dilatometer. For self healing study, cracks were made on the glass sample both by scratching it with the diamond scribe and creating a notch using diamond cutting blade. After cracks were made on the glass sample, it was placed inside a furnace equipped with a CCD camera. Video imaging of the cracks was monitored and recorded while it was heated at a rate of 5^0C/min to the desired temperature.

Seal making

YSZ was sealed to Crofer metal using glass 4. This was done by placing the glass between YSZ and Crofer and heating this assembly to 700^0C-900^0C. After seal was made it was first tested at room temperature to check, its hermeticity.

Seal testing

The seals were tested at room temperature and higher temperatures by pressurizing the seal from one side and monitoring the pressure of that side as a function of time using a pressure transducer. A hermetic seal shows a very little or no change in the pressure as a function of time. The seals were also tested after thermal cycling at different temperatures to assess the effect of thermal cycling on hermeticity. Seal was attached to steel housing with which a steel tube was connected. This assembly of seal and steel housing was placed inside a tube furnace with one closed end. Other end of the tube was sealed against a metal flange with openings for vacuum pump, thermocouple, gas entry into the furnace and gas entry into the steel housing assembly. This set up gave the ability to control the temperature and pressure of the furnace as well as the pressure inside the steel housing. By controlling the pressure in the furnace and steel housing one can control the □P experienced by the seal. Initial seal testing was done at room temperature for 20 hours at a □P ~ 150 Torr. Higher temperature leak tests were done for 1 hour at □P of 40 or 20 Torr.

RESULTS AND DISCUSSIONS

Thermal expansion characteristics of Crofer, glass 4, and YSZ are shown in Figure 1. Crofer expands more than glass and YSZ and hence it is under tensile stress after seal is cooled down from seal making temperature to room temperature. On the other hand glass is under compression, which is desirable. Glass becomes sufficiently fluid at a temperature above 700^0C, which enables it to flow and conform to Crofer and YSZ surfaces at seal making temperature. In order to flow, it is required that glass does not crystallizes too much at sealing making temperature, glass 4 remains amorphous after 2 hour exposure at 800^0C (same as seal making condition), which is confirmed by X-ray study and SEM photograph of glass sample sintered at 800^0C for 2 hours. A comparison of X-ray pattern of glass powder and sintered glass sample is shown in Figure 2.

Self healing study of the glass shows that glass starts to flow and heal at a temperature of 540^0C and healing is complete around 600^0C, which is in accordance with the thermal expansion data which tells that glass is sufficiently fluid at a temperature above 600^0C. During self healing test no load was applied on the glass sample. It takes nearly 5 minutes for crack to completely disappear, which indicates that the rate of healing is sufficiently rapid. Still photographs from the recorded video of the crack on the glass sample before and after healing are shown in Figure 3. It shows that crack is completely healed.

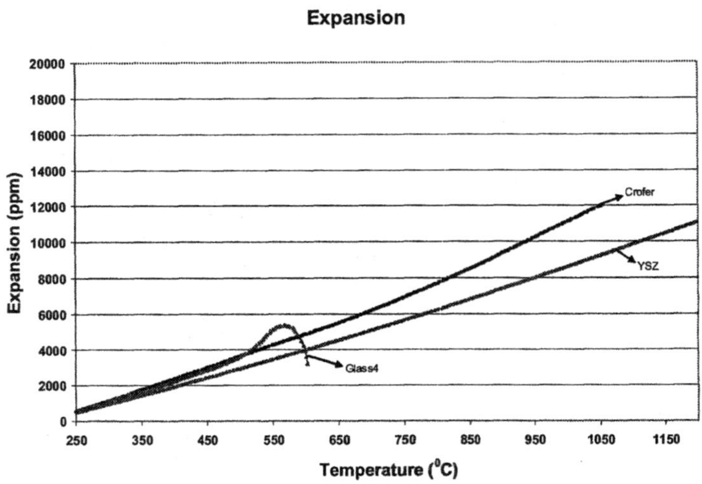

Figure1: Thermal expansion behavior of Crofer, glass 4 and YSZ

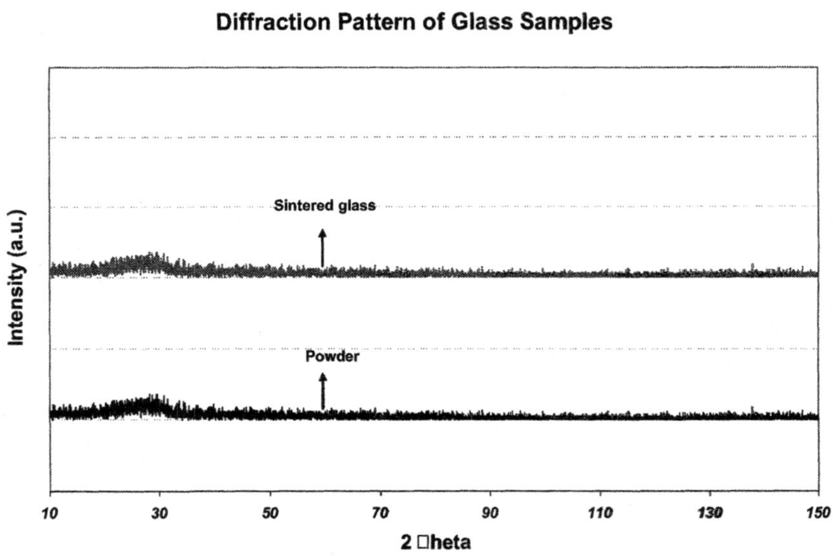

Figure 2: X-ray diffraction pattern of glass powder and a dense sintered glass sample.

Figure 3: Still photograph of crack on glass sample before and after healing

Seals of (Crofer- glass- YSZ) were successfully made by the process described above. These seals were tested at room temperature under a □P ~150 Torr. These seals were completely leak tight even after 24 hours of testing. Data on leak rate for one of the seal designated as seal #23 is shown in Figure 4.

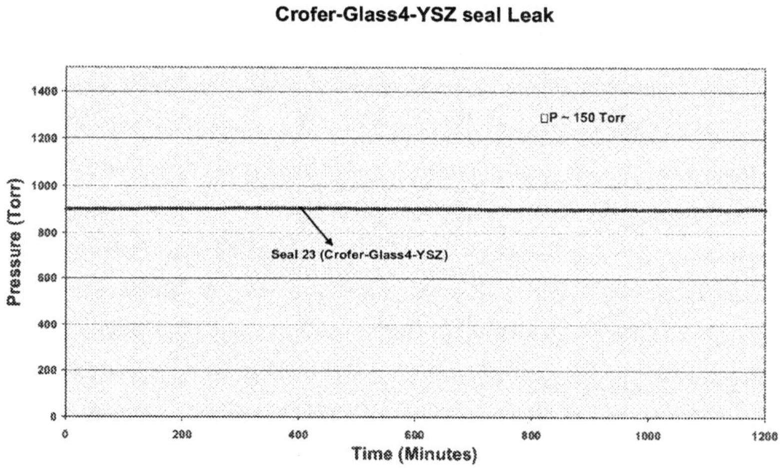

Figure 4: Leak check data of (Crofer-glass 4-YSZ) seals.

To check the viability of self healing of glass in real seal, a seal which was leaking at room temperature was kept at 800^0C for 2 hours and then cooled to room temperature. Then this seal was tested again at room temperature for leak check and it was found to be leak tight even after 24 hours of leak check under a □P of ~150 Torr. Figure 5, shows the data both before and after healing of this seal, which confirms that seal made with this glass can be healed at SOFCs operating conditions.

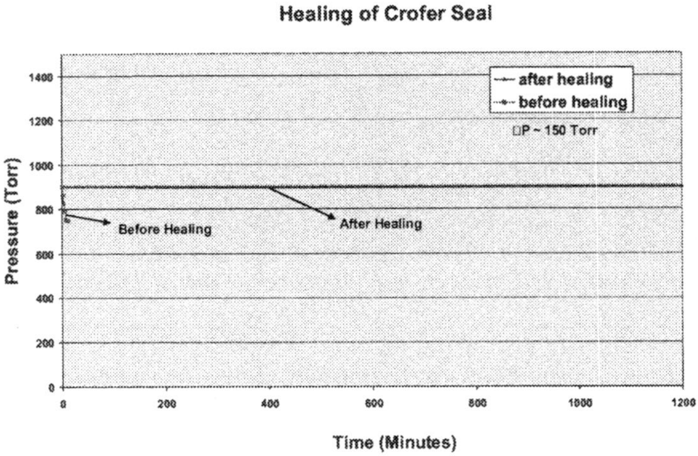

Figure5: Leak check data of seal before and after healing.

A seal designated as Seal 27 was tested for leak at higher temperatures during thermal cycling between room temperature and 800^0C. Data on leak rate of this seal at different temperatures during first thermal cycle is shown in Figure 6. Seal was completely leak tight at different temperatures.

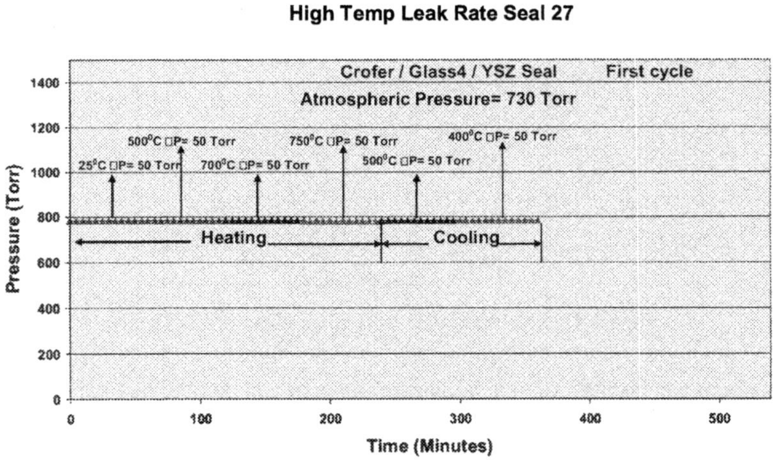

Figure 6: High temperatures Leak check data of the seal #27 during first thermal cycle.

Seal started leaking at 800^0C during third cycle. To heal the seal it was by kept at 800^0C for 30 minutes under no pressure difference from the atmosphere. After the healing seal became leak tight again, data of leak rate during third thermal cycle is shown in Figure 7.

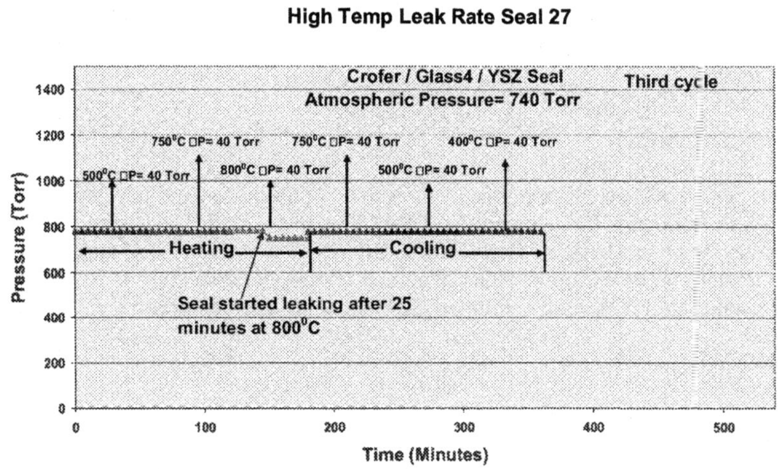

Figure 7: High temperatures Leak check data of the seal #27 during third thermal cycle.

Subsequent to the healing treatment, seal became leak tight and seal is leak tight even after 22 thermal cycles and testing of the seal is still going on. Healing of seal and its further hermetic behavior demonstrated that this seal can be repaired even if it starts leaking, just by keeping it at higher temperature for some time. Data of leak rate during thermal cycle#22 is shown in Figure 8.

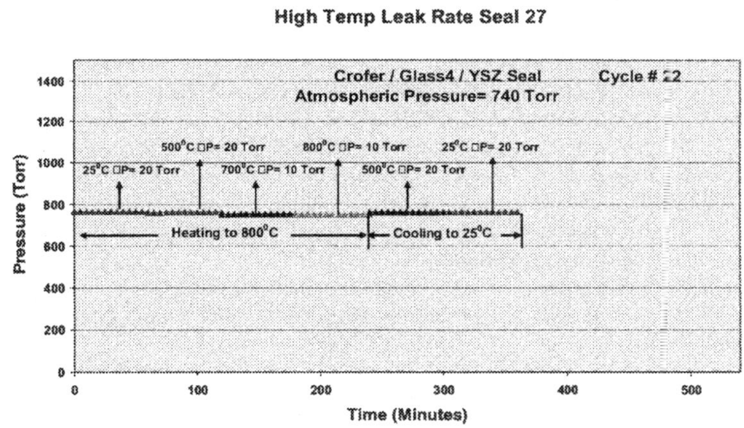

Figure 8: High temperatures Leak check data of the seal #27 during thermal cycle no.22.

CONCLUSIONS

Hermetic seals were made using a silicate glass for potential application in solid oxide fuel cells. Self healing behavior of the glass and the seals made using it were demonstrated. Stability of these seals under thermal cycling was also demonstrated as the glass seal remained leak tight even after 22 thermal cycles between room temperature and 800^0C. These results demonstrated the potential for self healing glass seals for application in SOFC.

ACKNOWLEDGEMENT

This project was partly supported by University of Cincinnati and US Department of Energy- SECA program through grant no.DE-FC26-04NT 42227. The guidance of Dr. Travis Shultz, Program Manager- Fuel Cells (DOE - NETL) is very much appreciated. Useful discussions with Drs. Don Collins, Wayne Surdoval, Prabhakar Singh, Matt Chou and Jeff Stevenson of PNNL and N. Minh of GE-Energy are also appreciated.

REFERENCES

1. N.Q. Minh, "Ceramic Fuel Cells," *Journal of the American Ceramic Society*, **76** [3] 563-88 (1993).

2. Hengyong Tu and Ulrich Stimming, "Advances,Ageing Mechanism and lifetime in Solid Oxide fuel Cells," *J. Pow. Source*, 127 284-293 (2004).

3. ZhenguoYang, Guanguang Xia, Kerry D. Meinhardt, K. Scott Weil, and Jeff W.Stevenson,_"Chemical stability of glass seal interface in intermediate temperature solid oxide fuel cell," *J.Mat.Eng.Perform.* 13 327-334 (2004).

4. S.B. Sohn, S.Y. Choi, G.H. Kim, H.S. Song, G.D. Kim, Stable sealing glass for planar solid oxide fuel cell, *J. Non Cryst Solids*, 297 103–112(2002).

5. Scott K. Weil, Christopher A Coyle, John S Hardy,Kim Jin Y Kim, Guan-Guang Xia, "Alternative Planar SOFC sealing Concepts," *Fuel cell bulletin*, 5 11-16 (2004).

6. N.H. Menzler, M. Bram, H.P. Buchkremer, and D Stöver,"Development of Gastight Sealing Material for Ceramic Components," *J.Eur.Ceram. Soc*, 23 445-454 (2003).

7. K. Eichler, G. Solow, P. Otschik, W. Schaffrath, "BAS (BaO–Al2O3–SiO2)-glasses for high temperature applications," *J. Eur. Ceram.Soc.* 19 1101–1104 (1999).

8. R. N. Singh, "High Temperature Seals for Solid Oxide Fuel Cells," *Proceedings of 29th International Conference on Advanced Ceramics and Composites*, Cocoa Beach, 25(3), 299-307 (2004).

9. R. N. Singh, "High Temperature Seals for Solid Oxide Fuel Cells," *proceedings of ASM international Conference*, Columbus, Oct 18-20, 2004.

10.R. N. Singh and S. S. Parihar, "Layered Composite Seals for Solid Oxide Fuel Cells," *Proceedings of 29th International Conference on Advanced Ceramics and Composites*, Cocoa Beach, (2005).

11. Prabhakar Singh and N. Q. Minh, Solid Oxide Fuel Cells: Technology Status," *Int. J. Appl. Ceram. Technol, 1 [1] 5-15 (2004.)*

12. P. Singh, "SOFC Seal Meeting SOFC Seal Meeting," SECA Core Technology Program,.Sandia National Laboratory, Albuquerque, July 8, 2003.

NOVEL SOL-GEL SYNTHESIS AND CHARACTERIZATION OF HIGH-SURFACE-AREA Pt-Ru CATALYSTS AS ANODES FOR DIRECT METHANOL FUEL CELLS

Moni Kanchan Datta, Jin Yong Kim[a] and Prashant N. Kumta
Department of Materials Science and Engineering, Carnegie Mellon University, Pittsburgh, PA 15213.
[a] Energy Materials, Materials Science Division, Pacific Northwest National Laboratory, Richland, WA 99352.

ABSTRACT

A novel complexation based sol-gel process has been developed to synthesize Pt-50at.% Ru based anode catalysts possessing high specific surface area with good catalytic activity. A homogeneous amorphous gel was successfully synthesized by complexing platinum(II) acetylacetonate and ruthenium(III) acetylacetonate with tetramethylammonium hydroxide (TMAH) used as a complexing agent. Phase pure Pt(Ru) solid solution of nominal composition Pt-50at.% Ru, possessing high specific surface area (~75m^2/g-140m^2/g) were successfully produced by controlled removal of carbonaceous species generated from the as-prepared precursor. The controlled removal of carbon related species has been successfully performed by controlled thermal treatment of the precursor using controlled oxidizing atmospheres such as 1% O_2 balanced with ultrahigh purity (UHP) argon. It has been identified that the nanocrystalline Pt(Ru) solid solution of nominal composition Pt-50at.% Ru possessing good chemical homogeneity with less amount of non-conducting carbon reveals an excellent catalytic activity, demonstrating the potential of the novel sol-gel process for synthesizing high-performance catalysts for direct methanol fuel cells.

INTRODUCTION

Of the different types of fuel cells, direct methanol fuel cells (DMFC) have become the focus of interest as suitable power sources for many applications [1-4]. The main advantage of DMFC is the use of liquid methanol rather than gaseous hydrogen as the operating fuel. Since liquid is much easier to store and handle than gaseous hydrogen, this is a significant advantage especially for mobile and portable applications. However, the currently achieved power densities of DMFC are still significantly below proton exchange membrane (PEM) and other fuel cell types. One of the main issues relate to the relatively slow kinetics of the electrochemical reactions occurring in the electrodes. Therefore, an important objective in the development of high-performance DMFC will be to improve the reaction kinetics of the Pt-50at.% Ru based anode by synthesizing high surface area Pt-50at.% Ru based catalyst. Several approaches have been reported [5-8] for synthesizing high surface area Pt-Ru catalysis by using different precursors of Pt and Ru, which require a follow-up washing treatment to remove the undesired residues. As a result, these processes are cumbersome and can cause significant loss of the starting noble metal precursors. Identification of an alternative process utilizing a non-halogen containing precursor could therefore be beneficial. In the present work, a new sol-gel based chemical process has been reported to synthesize Pt-Ru based complexes, which were thermally treated at various conditions to synthesize phase pure Pt(Ru) solid solution catalyst of nominal composition Pt-50at.% Ru possessing a high specific surface area with good catalytic activity.

EXPERIMENTAL

Phase pure and high surface area Pt-50at.% Ru based catalyst powders were synthesized using a complexation based sol-gel process. Platinum (II) acetylacetonate [Pt-acac: $Pt(C_5H_7O_2)_2$, Alfa Aesar] and ruthenium (III) acetylacetonate [Ru-acac, $Ru(C_5H_7O_2)_3$, Alfa Aesar] were used as the sources for Pt and Ru. Fig. 1 shows a schematic flow-chart of the procedure followed for this process. Pt-acac and Ru-acac of desired composition (Pt:Ru = 1:1) were dissolved at 50°C in 100 ml of acetone. Tetramethylammonium hydroxide [TMAH, $(CH_3)_4NOH$, 25% in methanol, Alfa], set at TMAH:(Pt+Ru) = 1.75:1, was then added to the solution to act as a high molecular weight organic complexing and hydrolyzing agent. Since Pt-acac and Ru-acac tend to phase separate during drying and evaporation of the solvent, the addition of an organic complexing agent such as tetramethylammonium hydroxide was found to be beneficial to yield homogeneous amorphous gels. After stirring for 10 min, the solvent was evaporated until the solution became viscous, transforming into a gel. The viscous gel was dried in air at 170°C for 10 h. The dried gels were then heat-treated in various conditions based on our earlier results [9] to synthesize high surface area Pt-Ru catalyst. It has been identified that a simultaneous control of the heat-treatment atmosphere, temperature and heat-treatment time can result in the formation of single-phase Pt-Ru catalyst exhibiting high specific surface area [9].

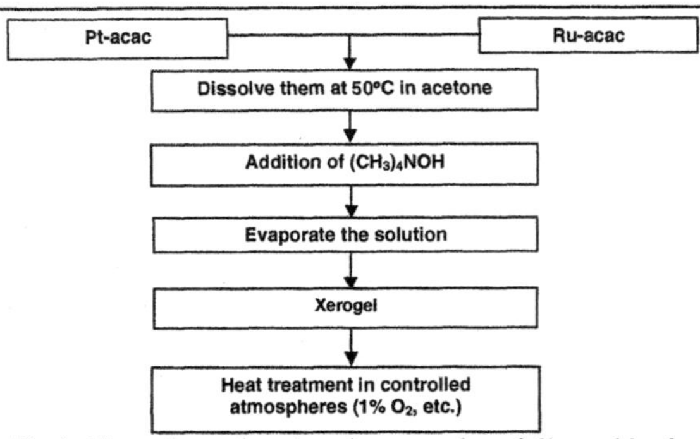

Fig.1. Flow sheet showing the procedure followed in the complexation based sol-gel process.

In order to identify the presence of phase/phases, X-ray diffraction (XRD, Philips XPERT PRO system with CuK_α radiation) study has been conducted on the as-prepared precursors as well as the heat-treated powders. The effective crystallite sizes of Pt-Ru catalyst were calculated from the most intense peak by Voigt function using single line approximation method [10]. Specific surface area of the catalyst has been measured using the Brunauer-Emmett-Teller (BET) technique (Autosorb 1, Quantachrome Instruments, Syosset, NY). In order to understand the phase formation and decomposition temperature, thermogravimetric and differential thermal analysis (TG/DTA) has been conducted on precursors as well as thermally treated powder using a simultaneous TGA/DTA machine

(TA 2960, TA Instrument, New Castle, DE). The TGA/DTA analysis has been carried out at a heating rate of 10°C/min from room temperature to 500°C in various atmospheres.

Electrochemical characterization was conducted on selected samples using a three-electrode cell. The working electrodes for electrochemical characterization were prepared by spreading the catalyst ink of the respective powders on teflonized carbon papers (applied on 1.5 cm^2 in area). The ink consists of 88 wt% Pt-Ru catalysts, and 12 wt% Nafion (5 wt% solution in mixed alcohols, Aldrich). The Pt loading on the electrode was $0.2mg/1.5cm^2$. The working electrodes were electrochemically tested using a Potentiostat (Model 553, ECO Instrument). A solution containing 1 M methanol and 0.5 M sulfuric acid was used as an electrolyte as well as a source of the fuel. A Mercury/Mercurous sulfate electrode that has a potential of +0.658V with respect to standard hydrogen electrode (SHE) was used as the reference electrode.

RESULTS AND DISCUSSION

The XRD patterns of the as prepared precursor powders derived using a molar ratio of TMAH:(Pt-acac + Ru-acac) = 1.75:1 and a nominal composition of Pt-acac:Ru-acac = 1:1 are shown in Fig. 2a. The as-prepared gel precursor exhibits an amorphous structure (Fig. 2a), which indicates its good homogeneity due to the possible complexation and hydrolysis of Pt-acac and Ru-acac with TMAH. This as-prepared

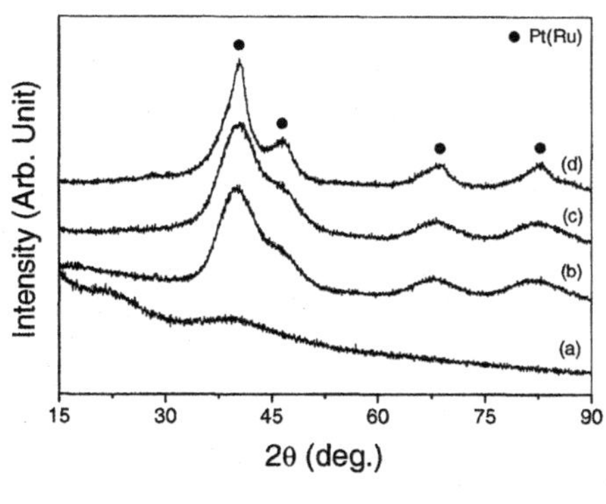

Fig.2. X-ray diffraction traces of the powders derived using tetramethylammonium hydroxide [TMAH : (Pt-acac+Ru-acac) = 1.25:1]; (a) as-prepared, (b) heat-treated in UHP-Ar at 500°C for 4h, (c) Sample A and (d) Sample B.

precursor was heat-treated in various conditions to synthesize nanocrystalline pure Pt(Ru) solid solution of nominal composition Pt-50at.% Ru. Based on the results of DTA/TGA of the precursor (Fig.3), which shows the decomposition of the precursor completed at around ~500°C, the as-prepared precursor was heat-treated at 500°C for 4h under UHP Ar atmosphere in order to synthesize nanocrystalline Pt-Ru based alloy. The XRD patterns of thermally treated powder (Fig. 2b) shows the formation of nanocrystalline Pt-Ru alloy with crystallite size ~6nm. However, the presence of large amount of non-conducting carbon has been detected based on the DTA/TGA analysis conducted on the thermally treated powder in air. The DTA/TGA traces (Fig. 4) performed under air shows a large exothermic reaction at 280°C with enormous weight loss (~60%) which is expected to

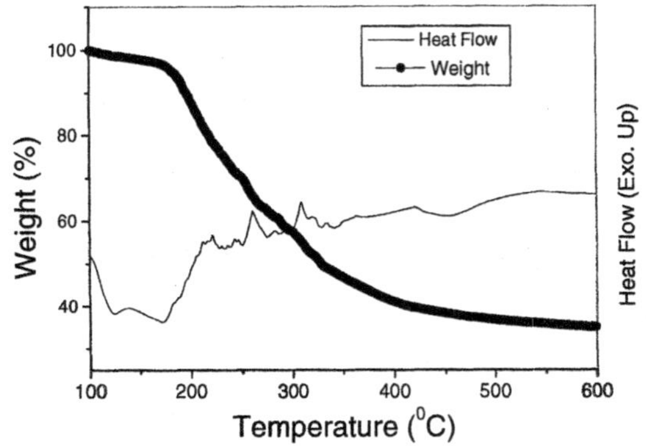

Fig.3. TGA/DTA traces conducted on the as-prepared precursor in Argon atmosphere.

arise due to the removal of carbonaceous species from the thermally treated powder. This thermally treated powder shows a specific surface area ~2m^2/g due to the presence of large amount of non-conducting carbon (~60 wt%).

In order to remove the carbon related species to form the desired nanocrystalline Pt(Ru) solid solution with high specific surface area, the thermally treated powder was

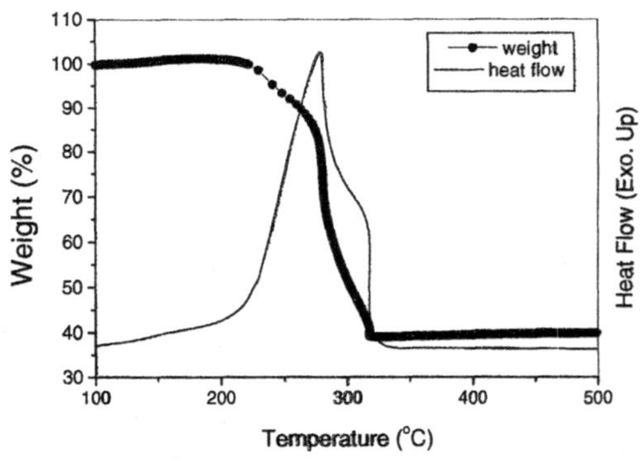

Fig.4. TGA/DTA traces conducted on the heat-treated powder under UHP-Ar at 500°C in air.

further thermally treated at 300°C for 4h in the presence of 1% O_2 balanced by UHP Ar. The oxygen in UHP Ar will serve to remove the carbon in a controlled fashion thus yielding the single phase Pt-Ru alloy. The XRD patterns of thermally treated powder at 300°C in Ar-1% O_2 (Sample A) shows the presence of nanocrystalline (~7nm) Pt-Ru alloy which suggest no significant grain growth occurs during thermal treatment at 300°C under Ar-1% O_2 atmosphere. The DTA/TGA analysis (Fig. 5) shows that the carbon content in Sample A is reduced to ~10 wt.% with respect to ~60wt.% in the sample thermally treated at 500°C at Ar atmosphere. This result clearly suggests that the thermal treatment conducted in the presence of 1% O_2 is helpful in removing the non-conducting carbon related species from the final product. Sample A shows a specific surface area of

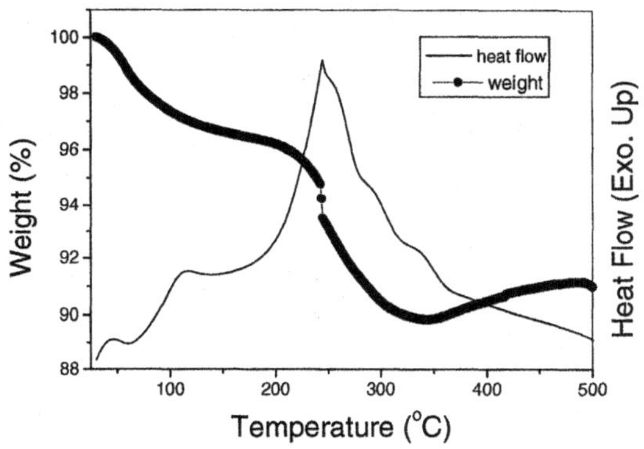

Fig.5. TGA/DTA traces conducted on Sample A in air.

~139m^2/g in comparison to 2m^2/g corresponding to the sample thermally treated at 500°C under Ar atmosphere. This result clearly suggests that the significant removal of non-conducting carbon, which facilities to form porous structure of Pt-Ru alloy, could enhance the specific surface area of the Pt-Ru alloy. The lattice parameter of nanocrystalline Pt-Ru alloy of Sample A, calculated from the most intense peak of Pt, corresponds to ~0.3904nm which is close to that of pure Pt [11]. This result suggests that the temperature and time of heat treatment is not sufficient to form a solid solution of Ru into Pt to form the nanocrystalline Pt(Ru) solid solution which is known to decrease the lattice parameter of Pt by the formation of a solid solution with Ru [11]. It is also possible that the presence of non-conducting carbon act as a diffusion barrier for the interdiffusion between Pt and Ru to form the solid solution of Ru with Pt.

In order to improve the alloying between Ru and Pt to form a solid solution in the nanocrystalline Sample A, which is probably a mixture of nanocrystalline Pt and Ru, Sample A was thermally treated further at 400°C for 2h in UHP Ar atmosphere. The XRD pattern of the thermally treated powder (Sample B), shown in Fig. 2b, shows that the Pt peaks are shifted towards higher 2θ value which suggests the formation of solid solution of Ru in Pt. The lattice parameter of Sample B, calculated from the most intense peak of Pt(Ru), correspond to ~0.3859nm which is very close to that corresponding to the composition of Pt-50at.% Ru solid solution (~0.3860nm). However, Sample B shows a specific surface area of ~75m^2/g, which is almost 45% lower than that of Sample A due to the significant grain growth of Pt(Ru) alloy at higher temperature. It is also interesting to note that there is significant reduction of the non-conducting carbon and presence of only 2% non-conducting carbon has been detected from DTA/TGA analysis (Fig. 6) in

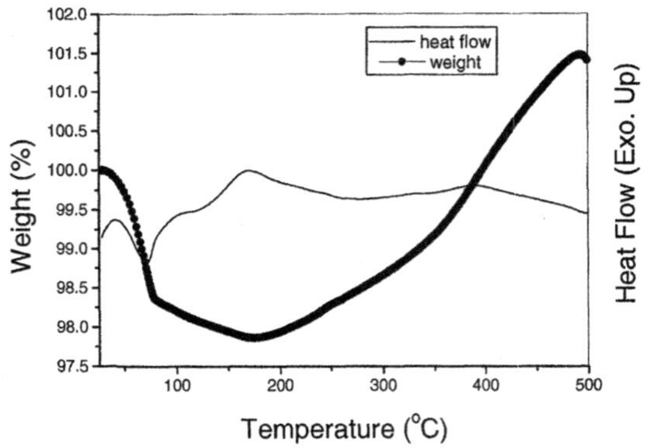

Fig.6. TGA/DTA traces conducted on Sample B in air.

Sample B which is ~75% lower than Sample A.

Results of electrochemical tests, i.e. polarization curve at 40°C and 60°C, conducted on sample A and Sample B is shown in Fig. 7 and 8, respectively. In order to compare the electrochemical performance of Sample A and Sample B with commercially available Johnson Matthey (JM) catalyst, the polarization curve of JM catalyst (specific

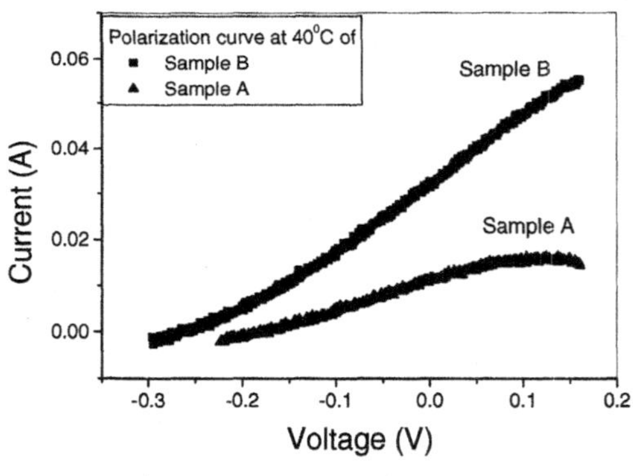

Fig.7. The polarization curves of Sample A and Sample B tested at 40°C.

surface area ~72m^2/g) has also been plotted in Fig. 8 tested at 60°C. At 40°C and 60°C, Sample A shows ~0.0045A/1.5cm^2 and ~0.009A/1.5cm^2 current density, respectively at –0.10V. Whereas, Sample B shows ~0.017A/1.5cm^2 and ~0.029A/1.5cm^2 current density, respectively at –0.10V, which corresponds to 70% improved performance in comparison to Sample A. On the other hand, Sample B shows comparable performance with commercial JM catalyst (~0.034A/1.5cm^2 current density at –0.10V) at 60°C. The low performance of Sample A in contrast to Sample B catalyst is probably due to the improper alloying between nanocrystalline Pt and Ru to form Pt(Ru) solid solution and

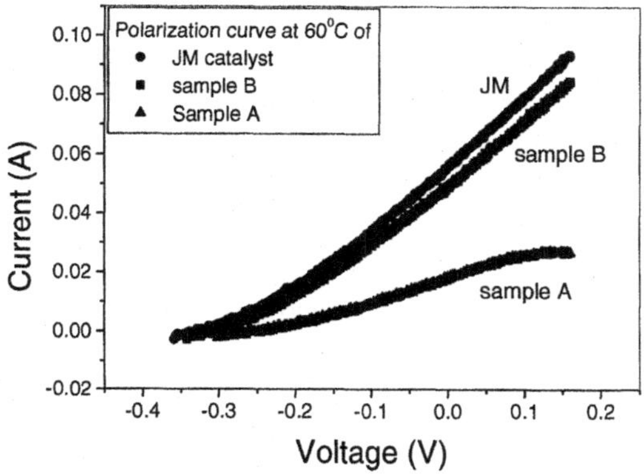

Fig. 8. The polarization curves of Sample A and Sample B along with JM catalyst tested at 60°C.

also the presence of 10wt.% non-conducting carbon. The above results suggest that the nanocrystalline Pt(Ru) solid solution of nominal composition Pt-50at.% Ru with good chemical homogeneity (perfect alloying) with less amount of non-conducting carbon reveals an excellent catalytic activity, demonstrating the potential of the novel complexation based sol-gel chemical process for synthesizing high-performance catalysts for direct methanol fuel cells.

CONCLUSION

A novel complexation based sol-gel process has been developed to synthesize Pt-Ru based anode catalysts possessing high specific surface area with good catalytic activity. A homogeneous gel was successfully synthesized by complexing Pt-acac and Ru-acac with TMAH using as a complexing agent. Phase-pure powders possessing high specific surface area 140 m^2/g (Sample A) was obtained by thermal treatment of gel precursor at 300°C for 5h using controlled oxidizing atmospheres such as 1% O_2 balanced with UHP Ar. However, the electrochemical properties of this material show poor performance due to the poor alloying between nanocrystalline Pt and Ru, and the presence of 10% non conducting carbon. Formation of Ru solid solution in Pt with a specific surface area ~75m^2/g (Sample B) has been obtained by conducting a subsequent thermal treatment of Sample A at 400°C for 2h in UHP Ar. The electrochemical properties of Sample B shows 70% improved performance in comparison to Sample A and comparable performance in relation to the commercially available JM catalyst. It has also been noticed that the current density of the catalyst is two times higher when the operating temperature is increased from 40°C to 60°C. The low performance of Sample A in contrast to Sample B catalyst is probably due to the improper alloying between nanocrystalline Pt and Ru to form a solid solution of Ru with Pt, and also the presence of 10wt.% non-conducting carbon. The above electrochemical test results demonstrate the potential of the novel complexation based sol-gel process utilizing non-halogen precursors for synthesizing high-performance catalysts for direct methanol fuel cells.

ACKNOWLEDGEMENTS

The authors acknowledge the financial support of DuPont Fuel Cells and the Pennsylvania Infrastructure Technology Alliance (PITA) for supporting this work. Technical assistance of Dr. Lin Wang of DuPont Fuel Cells is also acknowledged.

REFERENCES

[1]B. D. McNicol, D. A. J. Rand and K. R. Williams, *J. Power Sources*, **83,** 15 (1999).

[2]S. Surampudi, S.R. Narayanan, E. Vamos, H. Frank and G. Halpert, A. LaConti, J. Kosek, G.K. Surya Prakash, and G.A. Olah, *J. Power Sources*, **47**, 377 (1994).

[3]X. Ren, M.S. Wilson, and S. Gottesfeld, *J. Electrochem. Soc.,* **143**, L12 (1996).

[4]K. Y. Chan, J. Ding, J. Ren, S. Cheng and K. Y. Tsang, *J. Mater. Chem.*, **14,** 505 (2004).

[5]X. Zhang and K. Y. Chen, *Chem. Mater.*, **15**, 451 (2003).

[6] B. Yang, Q. Lu, Y. Wang, L. Zhuang, P. Liu, J. Wang and R. Wang, *Chem. Mater.*, **15**, 3552 (2003).

[7]Y. C. Liu, X. P. Qiu, Y. Q. Huang and W. T. Zhu, *J. Power Sources,* **111**, 160 (2002).

[8]Y. Takasu, T. Kawaguchi, W. Sugimoto and Y. Murakami, *Electrochimica Acta,.* **48**, 3861 (2003).

[9]Jin Yong Kim, Z.G. Yang, C.-C. Chang, T.I. Valdez, S.R. Narayanan, and P.N. Kumta, *J. Electrochemical Society*, **150,** A1421 (2003).

[10]Th. H. de Keijeer, J. I. Langford, E. I. Mittemeijer and A. B. P. Vogels, *J. Appl. Crystallogr.*, **15,** 308 (1982).

[11]M.-S. Lofflerm H. Natter, R. Hampelmann, K. Wippermann, *Electrochimica Acta,* **48** 3047 (2003).

GRAIN BOUNDARY SEGREGATION AND CONDUCTIVITY IN YTTRIA-STABILIZED ZIRCONIA

Monika Backhaus-Ricoult, Michael Badding and Yves Thibault*
Crystalline Materials Research or Characterization Science*, Science & Technology
Corning Incorporated
Corning NY 14831, USA

ABSTRACT

Grain boundary segregation of Y, Mn, La and Fe in doped and undoped 3YSZ and 8YSZ ceramics and its influence on the intra- and intergranular conductivity is studied by spatially resolved analytical transmission electron microscopy, electron energy loss spectroscopy and impedance spectroscopy. EELS profiles reveal significant enrichment of di- and trivalent ions in an 1-2nm wide grain boundary core with total concentrations of Mn^{2+}, Fe^{2+}, La^{3+} and Y^{3+} reaching up to 40cat%. The segregation factors depend on the crystallographic structure of the boundary. Enrichment of di- and trivalent dopants produces a negative grain boundary core charge. EELS scans also show the space charge layer. It extends over a few nanometers and is characterized by an increased O K absorption edge height compared to the grain core and characteristic O K ELNES features. O K edge height and fine structure directly indicate the oxygen vacancy depletion in the space charge layer and allow to track local changes in chemical bonding.

Compared to 3YSZ ceramics, cation segregation is less pronounced in 8YSZ ceramics.

Inter- and intragranular resistivity of doped and undoped ceramics are measured by impedance spectroscopy in the temperature range of 200-500°C. While the oxygen ion bulk conductivity of YSZ ceramics remains unaffected by the dopants, the grain boundary resistance suffers a significant increase. The behavior is explained by strong interactions of oxygen vacancies with di- and trivalent cations in the grain boundary core and by the lowered oxygen vacancy concentration in the space charge layer.

INTRODUCTION

Yttria stabilized zirconia and $(Sr,La)MO_3$ perovskites with M = Mn, Fe, Co, Ni are present material choices for solid oxide fuel cell (SOFC) electrolyte and cathode materials. Both, cubic 8YSZ and tetragonal 3YSZ are taken into consideration because of their respective advantages. The oxygen ion bulk conductivity in fully stabilized cubic zirconia is by a factor of two higher than that of tetragonal 3YSZ[1,2]. Tetragonal 3YSZ has superior strength and thermo-mechanical properties that enable processing of much thinner electrolyte layers[3]. In addition, 3YSZ ceramics exhibit less chemical reactivity with lanthanum strontium manganate than the cubic 8YSZ phase[4]. YSZ electrolyte materials are polycrystalline ceramics. Thus, their oxygen ion conductivity comprises bulk and grain boundary contributions. While the bulk conductivity is a direct function of the level of heterovalent dopants and the associated oxygen vacancy concentration[5], literature data on grain boundary conductivity present a broad scattering[1,2] that cannot be easily explained. In addition, the grain boundary resistance is usually two to three orders of magnitude higher than the intragranular resistance.

It shall be recalled that YSZ electrolytes are typically obtained by ceramic processing routes. Even the purest commercial starting powders still contain impurities at the level of 50-100ppm (silica, alumina, iron oxide, sodium oxide, hafnium oxide as typical impurities). As a consequence, even purest YSZ ceramic electrolytes have at least this level of impurities.

Low-purity YSZ ceramics usually contain much higher impurity levels. A high silica content is often responsible for the presence of second phase in such ceramics. Triple points are decorated by glass pockets and grain boundaries are more or less continuously wetted by glass films. Such glassy films have the effect of a blocking layer for oxygen ion conduction[2]. The very high grain boundary resistance in YSZ ceramics with high silica content can easily be explained by these observations. In very clean ceramics, no second phases or glassy grain boundary films are found; their grain boundary resistance, however, still remains by a factor of hundred to thousand higher than the intragranular resistance. In the literature, an intrinsic grain boundary blocking effect is suggested[6,7]; this assumption is supported by electrochemical measurements; however, no evidence for such effect is brought yet by direct microstructural observations.

A grain boundary can be considered as the transition zone between two crystals of different orientations with a relaxed structure and energy topography that promote local variations in charge distribution and chemical composition (segregation). As a consequence, at the grain boundary itself and in its proximity, stoichiometry and chemical composition may differ from the bulk crystals. Within the few atomic layers of the “grain boundary core”, the chemical composition changes abruptly and evolves then continuously over the more extended space charge layer. Ionic crystals usually contain low concentrations of statistically distributed point defects in the bulk. Segregation of charged defects and species to the grain boundary core yields an electric “core” charge, which, through the coupling with the associated electric field, imposes a redistribution of the charged defects in proximity of the boundary within the so-called space charge layer[8]. The thickness of the space charge layer is on the order of the Debye-Hueckel length. The concentration profiles of the charged species in the layer can be derived from the constancy of the electrochemical potential over the entire ionic crystal, including the space charge layer. Modeling of grain boundaries in yttrium doped zirconia suggests that the enrichment of di- and trivalent dopants in the grain boundary core is associated with a depletion of oxygen vacancies in the space charge layer. Since the dielectric constant of zirconia is small and the heterovalent dopant level in partially or fully stabilized zirconia high, space charge layers are expected to be very narrow. Electrochemical measurements suggest a grain boundary “electrochemical” width of several nanometers; HRTEM crystal structure analysis indicates a “structural” width of 1nm.

Ab initio calculations provide the exact atom and charge distribution for some very special grain boundaries, but the energy topography of the uncountable possible grain boundaries and the atom distribution especially in presence of high dopant concentrations with high segregation factors still remain unaddressed and leave numerous questions on the grain boundary conductivity of YSZ ceramics and their variability.

In order to contribute to the understanding of the grain boundary segregation in YSZ and its influence on the intra- and intergranular conductivity, we studied in the present work the grain boundary chemistry in doped and undoped polycrystalline tetragonal 3YSZ and cubic 8YSZ by spatially resolved analytical transmission electron microscopy and electron energy loss spectroscopy, quantified the segregation and used those findings to interpret bulk and grain boundary specific area resistance of those ceramics.

EXPERIMENTAL

YSZ electrolyte layers were processed by slip casting and firing 3YSZ and 8YSZ powders (YSZ with 3 and 8 mol % Y_2O_3 from Tosoh, Japan) to full density at 1430°C. After firing, 3YSZ ceramics have a uniform microstructure and a narrow grain size distribution with an average grain size of about 300 nm. According to X-ray diffraction, they are composed of tetragonal phase with negligible amount of monoclinic phase. After long time annealing in air at temperatures between 700°C and 1300°C, no significant phase transformation or increase in grain size was observed. At the same firing temperature, 8YSZ ceramics sinter to larger grain size, ranging from 2 to 5 μm. Impurity levels in the ceramics were close to the initial powder quality. Specifications by Tosoh indicate for the powders < 50 ppm Al_2O_3, < 20 ppm SiO_2, 70 ppm Fe_2O_3, 200 ppm Na_2O.

The YSZ ceramics were doped with lanthanum, manganese or iron by exposing them at 1250°C for different durations to $(La,Sr)MnO_3$ or $(La,Sr)FeO_3$. The dopant concentrations were determined by microprobe and ICP analysis. Global dopant levels in the samples are summarized in Table 1.

Table 1: Global dopant concentrations in studied YSZ ceramics

Sample	wt% Y_2O_3	wt% MnO	wt% FeO	wt% La_2O_3
3YSZ-0	6.5	-	-	-
3YSZ-A	6.5	0.35	-	0.2
3YSZ-B	6.5	0.45	-	0.3
8YSZ-C	18	-	0.5	0.3
8YSZ-0	18	-	-	-
8YSZ-A	18	0.03	-	0.02
8YSZ-C	18	-	0.05	0.03

Bulk and grain boundary resistivity of undoped and doped electrolyte were determined by impedance spectroscopy. For testing, a symmetric two-electrode, four-wire set up was chosen. Identical electrodes of porous LSM/3YSZ were sintered on both sides of the electrolyte and connected via a porous (Ag,Pd)-YSZ layer (indicated in the sample label as 1) or a Pt-Au-Pd (indicated in the sample label as 2) contact to silver leads. In the entire set up, potential sources of impurities, such as Pt-paste or low purity metal leads were carefully avoided. The used quality of silver was 99.999%; the Pt-Au-Pd quality was lower with Fe (0.2 at%), K (400 ppm), Na (150 ppm) and Si as main impurities. A well defined nominal active area of 1 cm^2 was achieved by a cross printing pattern of the electrodes. The test cells were introduced in a protective alumina tube in a tubular furnace and brought to the test temperature. Impedance data were acquired with a Solartron system (1260 Frequency Response Analyzer and 1287 Electrochemical Interface) with no applied bias at temperatures between 200 and 500°C in air. The frequency was varied from 1 MHz to 100 mHz. The amplitude applied between working and reference electrode was 30 mV. 10 points per decade of frequency were measured while scanning from the highest to the lowest frequency. Bulk, grain boundary and contact contributions to the impedance were fitted by an equivalent circuit comprising a parallel resistor and constant phase element for each observed arc. Constant phase elements were used in the modeling instead of simple capacitors because they describe better the real system with its depressed arcs.

The local grain boundary chemistry in undoped and doped 3YSZ and 8YSZ electrolyte was studied by spatially resolved transmission electron microscopy (TEM) with energy dispersive analysis (EDX) and electron energy loss spectroscopy (EELS). Sample cross-sections were mechanically polished to an electron transparent wedge and then ion-milled for some minutes to remove the mechanically damaged surface area. In the goal of achieving very high spatial resolution and resolving the grain boundary chemistry and electronic structure, the VG HB501 STEM13 of the Cornell CCMR facility with its probe size of 0.22 nm was used. It is equipped with a cold field emission gun and operated at 100 kV. A collection semi-angle of the spectrometer of 15mrad, an energy dispersion of 0.3 eV/ch and an emission current of 2.3 pA were the typical operation parameters. The EELS data were acquired by a CCD camera from the Parallel Electron Energy Loss Spectrometer (Gatan PEELS). Spectra of the Sr M, Y M, Zr M, Mn L, Fe L, O K, La L, Si L and Al L characteristic absorption edges were acquired. The evolution of the chemical composition across the grain boundaries was tracked by acquiring line spectra for the different edges while stepping the probe discontinuously across a grain boundary (aligned parallel to the beam). Spectra were explored manually. The characteristic elemental signals were extracted from the spectra by subtracting the background under the characteristic edges under use of a power law fit. Comparison of boundary spectra to those of the adjacent bulk phases was used to identify major changes in composition and in electronic structure at the grain boundaries. Chemical shift and near edge fine structural details were then used to identify in a "finger print"-type analysis valence states and bonding in the grain boundary core. Concentration profiles were corrected for the beam broadening within the foil. Reference spectra of known perovskite compositions were used to quantify the segregation profiles.

Overlap of Y and Zr M absorption edges, broad edges and few characteristic features in the M-edges made it difficult to identify and quantify modifications in the Y, Zr and Sr M edges at the grain boundaries. Only a slight hump at the energy location of the Y $M_{2,3}$ at 300 and 312eV was occasionally noticed when studying 3YSZ ceramics. Even though this feature was more pronounced in 8YSZ ceramics, its separation from the Zr M-edge remained difficult because the edge features change with sample thickness. The same difficulties applied to the Sr-M edge. Therefore, these elements were mainly tracked by EDX in the VG, but with a lower spatial resolution. Spatially resolved study of the Y L, Zr L edges was not possible in the VG501B because the small probe size, 100 KV operation voltage did not provide sufficient signal intensity for the >2000 eV range.

RESULTS

Mobility of Dopants in 3YSZ and 8YSZ ceramics

Figure 1 illustrates the microstructure of the 3YSZ and 8YSZ ceramics. 3YSZ ceramics exhibit a very homogeneous microstructure with narrow grain size distribution and an average grain size of 320nm, while grain sizes in the 8YSZ ceramic are by a factor ten larger and follow a much broader distribution in size and shape. Grain boundaries typically show little curvature.

Upon exposure to lanthanum strontium manganate or ferrate, lanthanum and manganese or iron dissolve into zirconia and diffuse into the ceramic. Overall dopant concentrations of the studied samples are summarized in Table 1.

Bulk diffusion of the cations was found to be rather slow in both, cubic and tetragonal phase. At 1250°C, bulk interdiffusion coefficients range for all studied dopants in the order of 10^{-17} cm^2/s. Grain boundary diffusion is by a factor of 10^6 faster. At 1250°C, grain boundary interdiffusion coefficients of $D_{gb} = 10^{-10.5}$ cm^2/s for manganese, $D_{gb} = 10^{-10.5}$ cm^2/s for iron and $D_{gb} = 10^{-11.3}$ cm^2/s for lanthanum were determined in 3YSZ. In 8YSZ, grain boundary diffusion coefficients were by a factor of two smaller. Data for bulk and grain boundary diffusion in 8YSZ and 3YSZ ceramics at various temperatures are summarized in Figure 2.

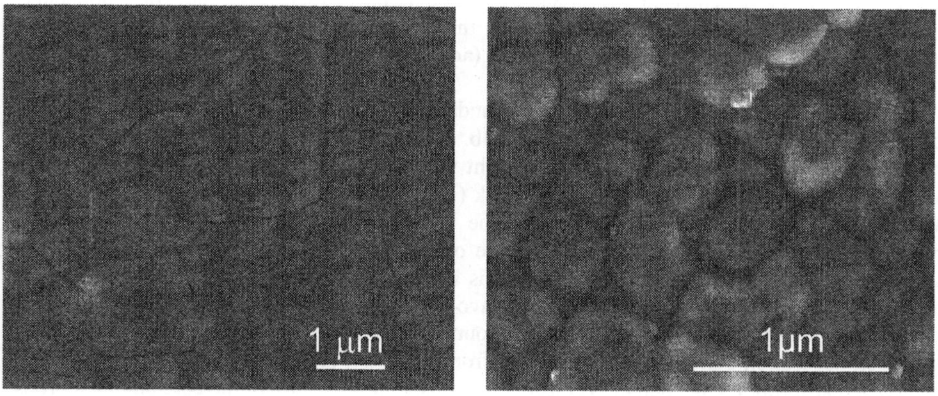

Figure 1: SEM micrographs illustrating the microstructure of the 3YSZ and 8YSZ ceramics

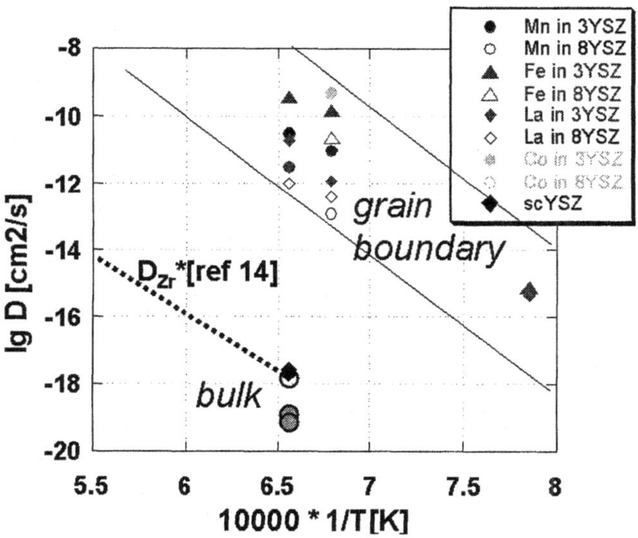

Figure 2: Bulk and grain boundary chemical diffusion coefficients of Mn, Fe, Co, La in 3YSZ, 8YSZ and single crystalline 10YSZ as a function of temperature (bulk diffusion coefficients at 1250°C are not further distinguished, bulk zirconium tracer diffusion coefficients from reference 14 are added for comparison as dotted line).

Grain Boundary Chemistry

EDX/TEM analysis of a large number of grain boundaries in the undoped electrolyte revealed significant segregation of yttrium to the grain boundaries. Typical spectra are presented in Figure 3a. The yttrium enrichment varies considerably between grain boundaries, showing that the core chemistry depends on the precise crystallographic structure of the grain boundary. Typically, the segregation factor for yttrium in the 3YSZ ceramic ranged between 2 and 3.5. Triple junctions exhibited slightly higher factors (around 3). Zr and Y M ELNES did not provide any information on changes in bonding.

However, the oxygen edge of the grain boundary core showed characteristic modifications compared to the adjacent bulk grains, Figure 3b. They result from yttrium segregation to the grain boundary core. Firstly, the O K edge height was found to be low in the core, indicating a lower local oxygen content compared to the bulk (low loss spectra did not reveal any change in sample thickness). A possible explanation is the lower atom density in the core region. This change goes hand in hand with a change in the oxygen near edge fine structure. The relative intensity of the peak doublet at 527/531eV is changed, and the doublet peak broadened, suggesting that additional electronic states are involved in the core bonding. Secondly, as shown in O K edge height evolution across the grain boundary in Figure 3c, a more continuous change in oxygen absorption edge height is observed from the grain boundary core towards the bulk. The oxygen absorption edge height is considerably raised in proximity of the core and then drops slowly over several nanometers to the bulk characteristic height. The oxygen absorption edge height is directly proportional to the number of oxygen atoms in the analyzed volume[9]. For constant sample thickness, changes in O K edge height indicate changes in the oxygen atomic concentration and, therefore, can be interpreted in the case of the doped YSZ in terms of oxygen vacancy concentrations. The change in oxygen edge height then can be interpreted in terms of a severe depletion in oxygen vacancies in a space charge layer that extends over some nanometers.

The ceramics contain a low level of silica and alumina impurities. Conventional and high resolution TEM analysis did not reveal any discontinuous glass films or triple point glass pockets. However, these impurities may be enriched at grain boundaries. For further insight in their segregation, the absorption edges of those elements were tracked and monitored while scanning across numerous grain boundaries. Effectively, by averaging over multiple acquisitions, it was possible to detect a very weak aluminum signal in the bulk and at the grain boundary. The absolute aluminum concentrations were very small, the enrichment factor at grain boundaries less than 1.5. Within the detection limit of the technique (5-10% of a monolayer), silicon was not observed. The low silicon bulk concentration (confirmed by microprobe analysis) was not accessible by the technique. At grain boundaries, no silicon was detected. It is concluded that no significant silicon segregation to grain boundaries occurs in the studied high purity 3YSZ. Further on, the larger size hafnium ions did not segregate to the boundaries.

Undoped 8YSZ electrolyte showed much less yttrium enrichment of the boundaries.

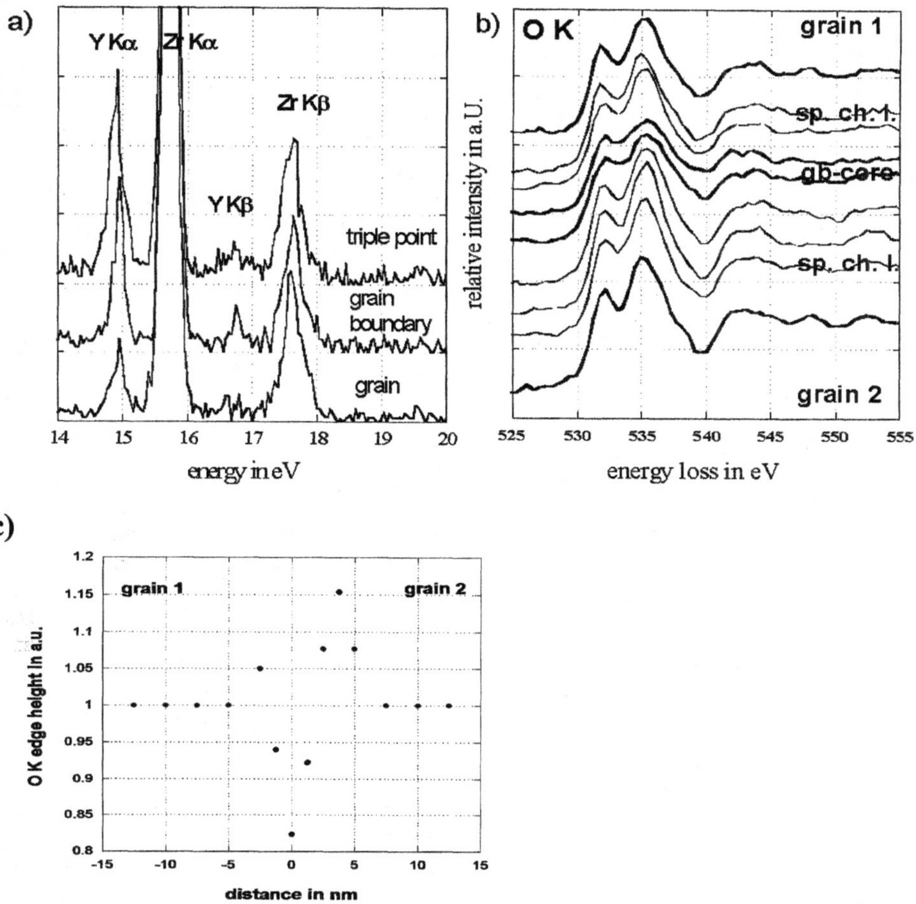

Figure 3: a) Normalized EDX/TEM spectra of the Y K and Zr K peaks at bulk grain, grain boundary and triple point locations in undoped 3YSZ ceramic showing the segregation of yttrium to grain boundaries b) O K ELNES line-scan across a grain boundary in the 3YSZ ceramic illustrating the particular fine structure in the core (gb-core) and increased oxygen edge height and doublet intensity ration in the space charge layer (sp.ch.l.). The distance between the subsequent spectra is 1.2nm. c) O K absorption edge height (in arbitrary units) evolution across the grain boundary

Doped electrolyte showed significant segregation of the di- and trivalen dopants to its grain boundaries. EDX/TEM spectra reveal the presence of Mn, La, Fe at the grain boundaries, and do not show any signal for these elements in the adjacent bulk grains. EELS/STEM investigations provided details on the grain boundary core composition. Effectively, high local concentrations of manganese and lanthanum or iron and lanthanum were detected in the grain boundary core. Segregations was found to be more important for grain boundaries of 3YSZ ceramics; 8YSZ ceramics showed more higher solubility of the dopants in the bulk grains and only a small grain boundary enrichment compared to bulk concentrations. 3YSZ ceramics showed negligible dopant concentrations in the bulk, while grain boundary core concentrations were in the order of 20-30 cat%. The segregation factor reached several orders of magnitude!

Figure 4 shows the absorption edges of manganese, lanthanum and oxygen in the grain boundary core and in the adjacent grains of a doped 3YSZ ceramic. Within the detection limit of the EELS technique, no manganese or lanthanum was found in the bulk 3YSZ grains. At the grain boundary, in contrast, strong signals of manganese (Mn $L_{2,3}$ at 640eV) and lanthanum (La $M_{4,5}$ at 840eV) are visible, Figure 4. In addition, the oxygen signal at the grain boundary is modified: In the grain boundary core, a change in the relative peak intensities of the doublet at 536eV was observed that reveals strongly defective zirconia. In the layer adjacent to the core, the doublet exhibits a changed intensity with a high second peak, characteristic for YSZ with low dopant levels and thus low oxygen vacancy concentrations. The oxygen absorption edge height evolves over several nanometers before stabilizing on the bulk value.

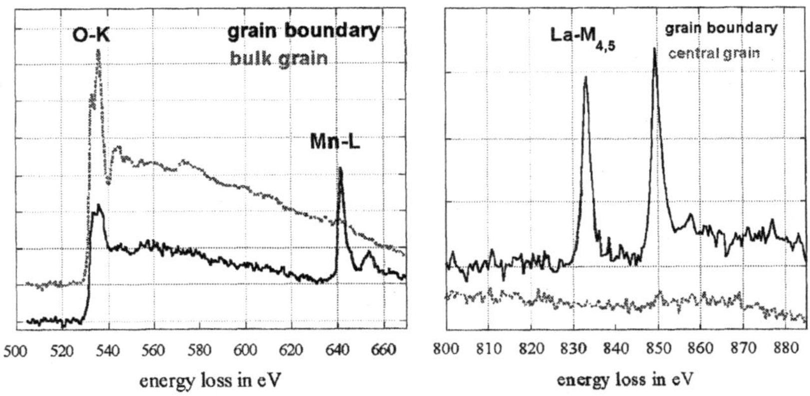

Figure 4: Absorption edges of oxygen, manganese and lanthanum in a bulk 3YSZ grain (black) and on the grain boundary core (dotted grey) in the 3YSZ-A ceramic

Numerous grain boundaries have been studied in samples with different dopant levels after their annealing in air at 1250°C. No modifications compared to the bulk were observed in Zr M, Y M and Sr M edges; however, it shall be recalled that these M edges are not sensitive to local bonding details and that minor changes cannot be identified. For manganese, iron and lanthanum, the height of the characteristic absorption edges evolves across the grain boundaries as demonstrated in Figures 5 and 6. A steep intensity profile was found with a maximum in the grain boundary core. The entire profile never extended over more than 2-3nm. The absorption edge height was quantified using bulk reference materials (La,Sr)MnO_3 and (La,Sr)FeO_3. Cation concentration profiles were deduced from the line spectra. The maximum concentration varied from grain boundary to grain boundary, reflecting a dependency on the precise crystallographic nature of the grain boundary. For manganese, concentration maxima varied between 10 and 25 cat%, for lanthanum, 9-12 cat%. The profiles were rarely symmetric, thus accounting for the crystallographic asymmetry of the boundaries.

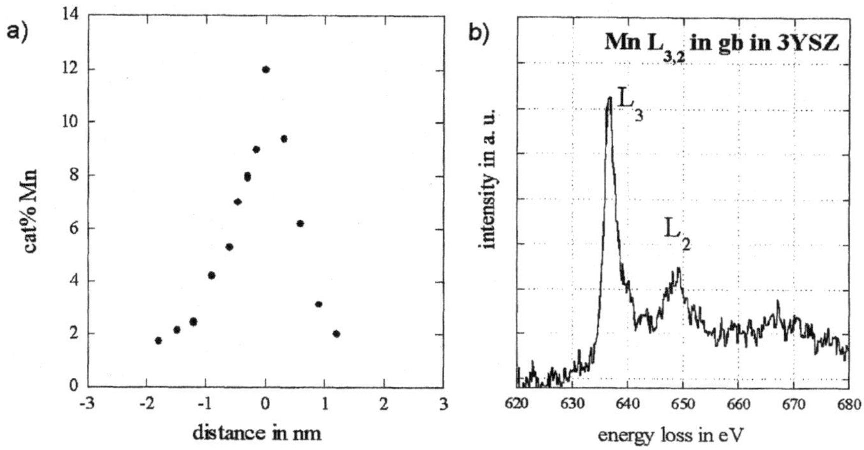

Figure 5: a) Series of manganese absorption edges acquired while crossing a with an interstep of 0.3nm, a) manganese concentration profile at a 3YSZ-A grain boundary obtained from Mn -scan across a grain boundary (position 0 nm), b) ELNES fine structure of the manganese absorption edge in the grain boundary core

The manganese valence state was determined from the $L_{2,3}$ chemical shift and the L_3/L_2 ratio. According to the literature, typical L_3/L_2 white line ratios are 5 for Mn^{4+} , 4.7 for Mn^{3+} and 3.4 for Mn^{2+} [10]. Manganese in the grain boundary core of 3YSZ, see Figure 5b, has a L_3/L_2 ratio of about 3, thus clearly indicating a main oxidation state of 2+ for manganese in the grain boundary core. Energy location and energy split confirm this statement.

Oxygen scans across the grain boundaries showed typically a decrease in absorption edge height of about 20% in the core as demonstrated in Figure 7. This corresponds well to a Mn^{2+} and La^{3+}-rich grain boundary core. For the core absorption edge, in addition, a change in the near edge fine structure was observed. The peak intensity was increased on the high energy side, yielding a high energy shoulder of the high energy doublet peak and an increased intensity in the doublet valley. Heavily doped defective zirconia in fluorite structure and pyrochlore usually show such a symmetric doublet; it can be speculated that a doublet with equal peak intensities might be related to oxygen vacancy ordering and deformation in the next neighbor symmetry. The high energy shoulder seems to be related to Mn-O-bonding.

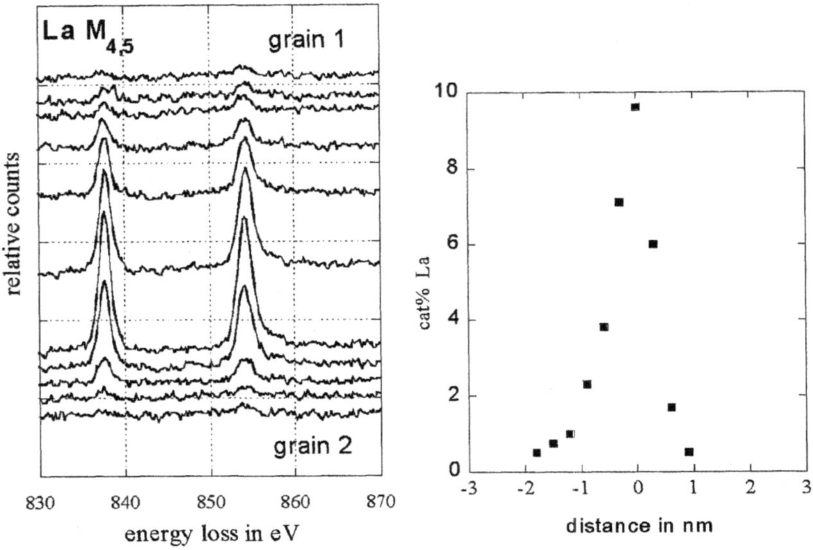

Figure 6: a) Series of lanthanum absorption edges acquired while crossing a grain boundary in 3YSZ-A with an interstep of 0.3 nm ; b) corresponding concentration profile of lanthanum across the grain boundary

Impedance Results

Typical impedance spectra for undoped and doped 3YSZ and 8YSZ ceramics are presented in Figure 9. The impedance spectra are composed of three arcs, a low frequency arc representing the bulk conductivity, an intermediate frequency arc representing the grain boundary resistivity and a more or less visible high frequency arc that originates from the resistance of the two electrolyte/electrode contacts. Comparison of the low frequency arcs of undoped and doped 3YSZ ceramics shows that the intragranular resistance does not significantly differ. The same is valid for 8YSZ ceramics. However, bulk conductivities of 3YSZ and 8YSZ are different. The impedance curves are fitted by an equivalent circuit[11]. The area specific resistance for the bulk R_g is computed from $\rho_g = R_g \times S/h$ with S being the measured area, h the electrolyte thickness (typically 20 µm for 3YSZ and 40 µm for 8YSZ ceramics). For pure 3YSZ at 300°C, a grain resistance of 147 kΩcm is obtained, for pure 8YSZ 386 kΩcm. Results of the activation energies are summarized in Table 3.

The second arc of the impedance spectra strongly changes with doping. Figure 9a demonstrates how the grain boundary impedance changes with dopant concentration. The grain boundary area specific resistance is derived from $\rho_{gb} = R_{gb} \times S_{gb}/n\ \delta_{gb}$ with δ_{gb} being the grain boundary core width, n the number of grain boundaries in the foil. Resistance data for bulk and grain boundaries are presented for a wide temperature range in Figure 10.

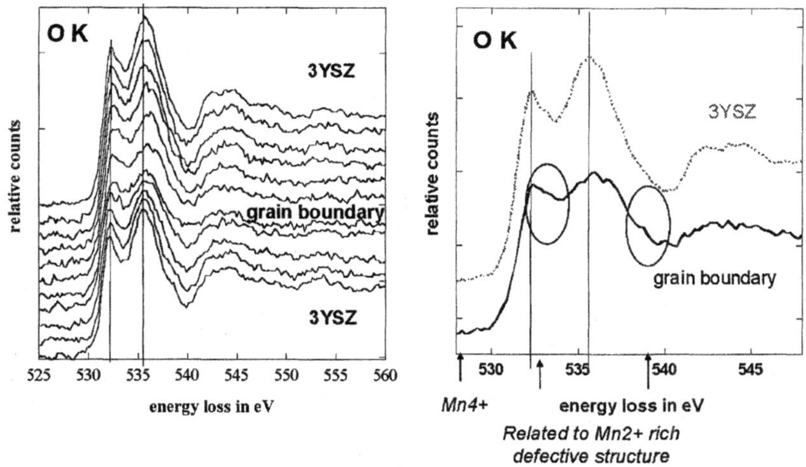

Figure 7: Series of O K absorption edges acquired while crossing a grain boundary in 3YSZ-A with 0.3nm as interstep; details of O K ELNES in bulk and grain boundary core

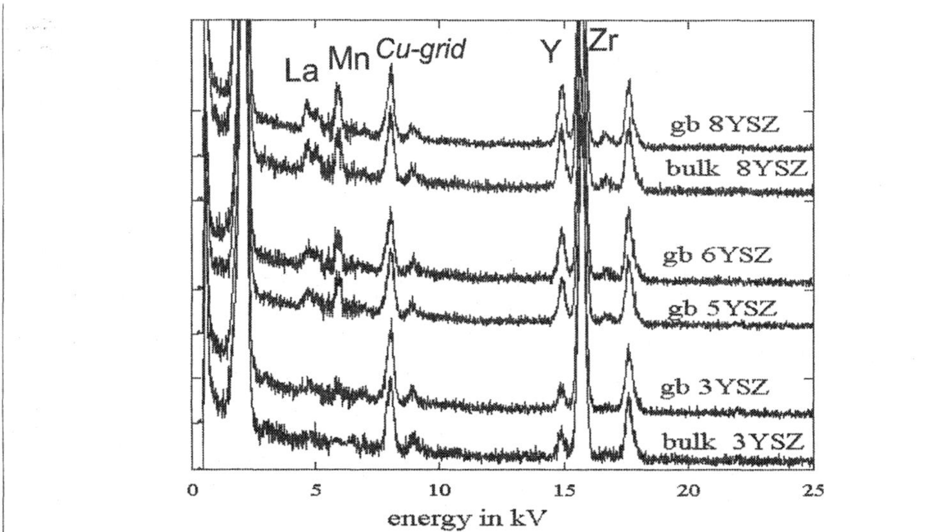

Fig. 8: Bulk and grain boundary EDX spectra in Mn/La-doped ceramics with different Y-content

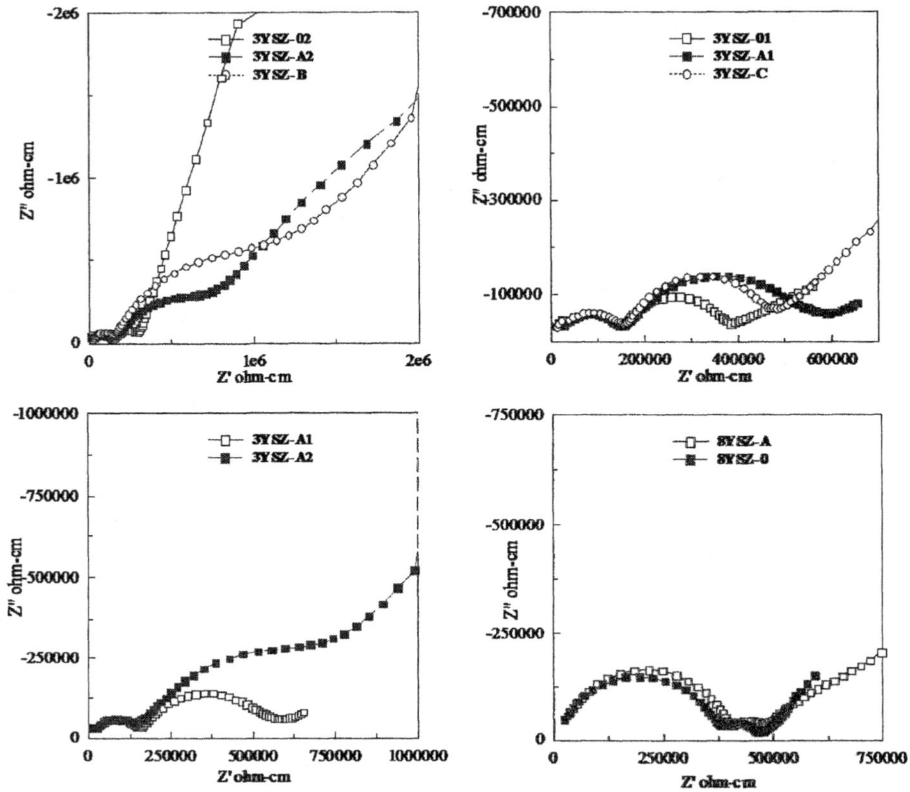

Figure 9. Typical impedance spectra of doped and undoped 3YSZ ceramic at 300°C: (a) comparison of 3YSZ electrolytes with different La/Mn-concentrations; (b) comparison of Mn/La- and Fe/La- doped 3YSZ; (c) effect of contact paste on undoped 3YSZ; d) comparison of undoped and Mn/La-doped 8YSZ. Global sample compositions are identified in Table 3.

For high purity materials, bulk and grain boundary resistance are of similar size. For 3YSZ materials, for temperatures below 300°C, usually the grain boundary resistance is smaller, for temperatures above 300°C, it is inversed. Highly doped ceramics like 3YSZ-B exhibit a dramatic increase in grain boundary resistance compared to the undoped material by a factor of 50 or larger. For 3YSZ-A1 and 3YSZ-C, no difference is noticed even though the overall concentration of iron and lanthanum is higher than the manganese and lanthanum concentration. This observation has to be explained either by a higher valence state of iron in the grain boundary core than manganese or by stability differences of the oxygen vacancy-cation pairs formed with manganese and iron. Since our EELS studies did not reveal significant differences in valence state, the letter explanation seems to apply. For 8YSZ ceramics, the grain boundary resistance is only little affected by dopants. This was expected based on the grain boundary core compositions in 8YSZ ceramics

Table 3: Activation energies for grain and grain boundary resistance in the YSZ materials

Sample	contact	E_A grain in eV	E_A gb in eV
3YSZ-02	Pt-Au-Pd	0.95	1.08
3YSZ-A1	Ag-Pd	0.90	1.19
3YSZ-A2	Pt-Au-Pd	0.91	1.10
3YSZ-B2	Pt-Au-Pd	0.92	0.96
3YSZ-B1	Ag-Pd	0.92	1.12
8YSZ-01	Ag-Pd	1.14	1.25
8YSZ-A1	Ag-Pd	1.09	1.2

DISCUSSION

Yttria-stabilized zirconia is an oxygen ion conductor with conductivities in the order of 2 10^{-2} Scm^{-1} and 28 10^{-2} Scm^{-1} at 1000°C for cubic 8YSZ and tetragonal 3YSZ, respectively. Cation diffusion is orders of magnitude slower[12,13,14] than oxygen ion diffusion. According to the literature[15], Zr* tracer diffusion is by a factor 10 faster in 3YSZ compared to 10YSZ with absolute values around $10^{-17}cm^2/s$ at 1200°C[16]. Diffusion of other dopant cations in YSZ scales with their cation radius; La^{4+}, Mg^{2+}, Ca^{2+}, Mn^{2+} and Ti show mobilities that are by a factor 3-7 higher than the Zr mobility [17,18,19,20].

During SOFC processing, after few hours of exposure to the cathode materials at 1200-1300°C, diffusion of manganese into the YSZ electrolyte was noticed and reported in the literature. We determined interdiffusion coefficients in LSM/YSZ and LSF/YSZ couples as a function of Y-level and temperature and distinguished bulk and grain boundary diffusion[21]. Bulk diffusion coefficients are in the order of $10^{-17}cm^2/s$ at 1250°C. Grain boundary diffusion is faster by a factor of 10^6. Both, bulk and grain boundary cation chemical diffusivities are by a factor three to ten faster in 3YSZ than in 8YSZ. High activation energies for both bulk and grain boundary diffusion yield a steep diffusion onset with temperature. Thus, considerable bulk concentrations of dopants or pollutants are obtained by diffusion only at temperatures above 1400°C. Grain boundary diffusion in 8YSZ and 3YSZ ceramics is already activated at temperatures as low as 1200°C and is responsible for fast doping of grain boundaries at such low temperatures.

It is generally agreed that di- and trivalent dopants introduce oxygen vacancies in the zirconia crystal lattice according to

$$M_2O_3 = 2M'_{Zr} + 3O^x_O + V_O \quad \text{and} \quad MeO = Me''_{Zr} + O^x_O + V_O$$

Fully stabilized zirconia adopts a deformed fluorite structure and requires for its stabilization about 7 cat% oxygen vacancies. Partially stabilized zirconia is a two phase mixture of cubic zironia with precipitates of tetragonal phase, which provide high fracture toughness by the martensitic transformation that they undergo under volume increase in the crack tip stress field. It was demonstrated that single phase very fine-grained tetragonal zirconia exhibits similar

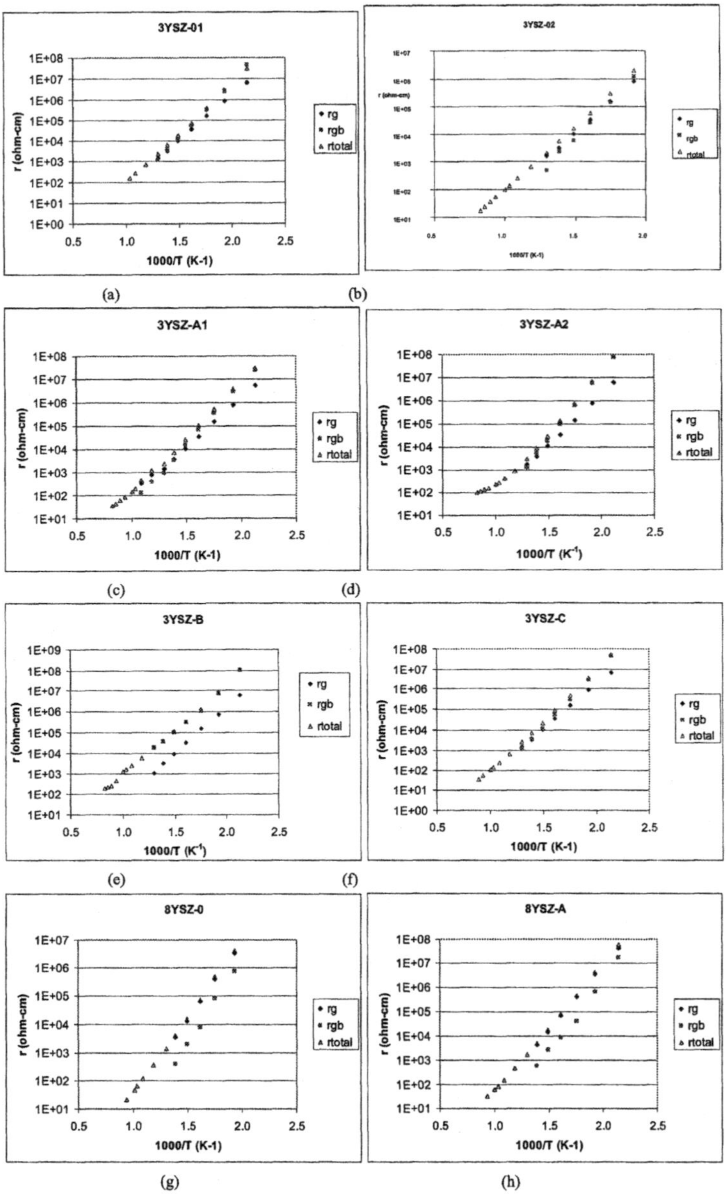

Figure 10. Arrhenius plots for grain and grain boundary resistance for YSZ ceramics as identified in Table 3.

properties as partially stabilized zirconia[3]. In bulk zirconia, the oxygen vacancy concentration increases with the dopant level, yielding the well-known relationship between dopant concentration and conductivity for the low dopant concentration range

$$\sigma = const./T\ [V_O^{\cdot\cdot}]\ ([V_O^{\cdot\cdot}]^{-1})\ e^{-EA/RT}$$

with R being the gas constant, T the temperature, EA the activation energy[22]. For high dopant concentrations, interactions between the defects introduce deviations from the ideal behavior. Usually the formation of defect pairs of type $\{Y_{Zr}'V_O^{\cdot\cdot}\}^{\cdot}$ is considered, but larger defect associates have also been suggested.

Two- and trivalent dopants segregate to the grain boundary core due to charge and size effects. Dopant and yttrium cations are mobile at temperatures above 1200°C in YSZ; at lower temperature, the dopant concentration profiles are frozen and only oxygen ions and anionic defects equilibrate. For the present case of zirconia, the space charge theory[6] predicts an extremely thin space charge layer due to high dopant concentrations (7-20 cat%) and a value of the dielectric constant of around 60.

Assuming that Mn^{2+} La^{3+}, Fe^{2+} and Y^{3+} are the dominating dopant species in the zirconia solid solution, the defect chemistry of the solid solution $(Zr,Y,Mn,La,Fe)O_2$ can be described by the following equations:

$$[VO^{\cdot\cdot}] = \tfrac{1}{2}[Y_{zr}'] + \tfrac{1}{2}[La_{zr}'] + [Mn_{zr}''] + [Fe_{zr}'']$$

Di- and trivalent dopants segregate to the grain boundary because of their charge and size; dominant driving force is the elastic energy gain. Segregation produces a local negative charge in the core and stabilizes cation vacancy and oxygen vacancy concentrations in the adjacent areas of the grains at levels different from the bulk equilibrium concentrations. The Boltzmann-Poisson equation indicates the evolution of their concentrations. Calculations of the defect concentrations in the space charge layer have been realized by Guo and Maier[6].

At the high dopant levels of 3YSZ or 8YSZ, the space charge layer shows a depletion in oxygen vacancies (cation vacancy concentrations are negligible in the bulk and in the space charge layer). Absolute oxygen vacancy concentrations and width of the space charge layer depend on the dopant concentration. In 8YSZ, the bulk oxygen vacancy concentration is around 20%. In the space charge layer close to the interface, the oxygen vacancy concentration is dropped to almost zero, but raises rapidly within a few atomic layers to almost the bulk value. In 3YSZ, the bulk oxygen vacancy concentration is smaller and the oxygen vacancy depletion close to the interface extends slightly further into the bulk. Our experimental observations suggest an extension of the space charge layer over 3-5nm.

In a simplified vision, the grain boundary can be described by a set of double layers with a heavily substituted negatively charged core and its surrounding severely oxygen vacancy depleted space charge layers. All three layers have to be considered as blocking for oxygen ion transport due to the strong interaction of di- and trivalent cations in the core with the oxygen vacancies and due to the severe depletion in oxygen vacancies in the space charge layers. For simplified impedance modeling, grain boundaries can be described as a single layer with very low oxygen ion conductivity.

Grain Boundary Chemistry In Undoped YSZ Ceramics

Due to its sensitivity to the local environment of the oxygen atom in the zirconia polymorphs, O K ELNES differs for cubic, tetragonal and monoclinic phase[23] and thus allows to distinguish the phases by the respective peak intensities of the doublet at 535eV. Several spatially resolved studies by TEM/EDX or TEM/EELS of grain boundaries in cubic and tetragonal zirconia are reported. In pure materials, yttrium is found to be enriched at grain boundaries. Segregation factors of up to 4 are reported in 3YSZ [22] (densified at 1450°C), and of 1.5 for 8YSZ [24]. These findings are in agreement with our observations of pure YSZ ceramics. We found a similarly small enrichment in 8YSZ boundaries (factor 1.2-1.6) and a larger enrichment in 3YSZ boundaries (factor 2-3.5). In the above mentioned literature, no modification in O K ELNES was noticed. This could be related to missing spatial resolution. We observed for 3YSZ grain boundaries a systematic evolution of the intensity ratio in the peak doublet at 535eV across the boundary with an increased peak intensity ratio at the boundary that evolved within a few nanometers into the typical 3YSZ bulk ratio. In terms of oxygen vacancy concentrations, our observation relates to the oxygen vacancy depleted space charge layer with its high depletion at the boundary that disappears over 3-5nm towards the bulk. In addition, we observed a drastic change of the OK ELNES in the grain boundary core; core OK ELNES exhibits the feature of highly doped defective cubic zirconia. This is not surprising. The core composition is 10YSZ or even more Y-rich.

For grain boundaries in undoped 8YSZ, we did not notice significant changes in O K ELNES at the grain boundary.

Grain Boundary Chemistry in Doped YSZ Ceramics

It is reported in the literature that yttrium segregation is easily suppressed in presence of other dopants such as aluminum. From our study of doped YSZ ceramics, we confirm that yttrium easily leaves grain boundary core sites to other dopants. Manganese, lanthanum and iron are present at much lower overall concentrations in the 3YSZ ceramic, but segregate heavily to the grain boundaries because of the larger strain and charge effects. Boundary cores in 3YSZ exhibit extremely high concentrations of di- and trivalent cations, for example 18 cat% manganese, 12 cat% lanthanum and 7 cat% yttrium. Such compositions are close to the bulk stability limit for cubic zirconia! Corresponding to the local environment of oxygen atoms in the grain boundary core, we detected in the grain boundary core O K ELNES features that match highly doped defective cubic zirconia with its height-equivalent O K doublet. Additional high and low energy shoulders in the core O K ELNES indicate the presence of electronic states that imply more direct binding contributions between Mn^{2+} and oxygen.

Grain Boundary Chemistry And Conductivity

Our impedance measurements indicate that the bulk oxygen ion conductivity is little affected by the presence of the dopants. This is not further surprising since almost the entity of the Mn, La, Fe dopants is segregated to the grain boundary and the grain composition is only determined by the yttrium content. The yttrium concentration determines the bulk conductivity; as clearly shown by the differences between 3YSZ and 8YSZ ceramics.

The grain boundary resistance in YSZ ceramics increases with the overall concentration of di- and trivalent dopants and thus with the local segregant concentration in the grain boundary core. If we consider grain boundary core compositions as determined by EELS of 18 cat% Mn, 12 cat% La and 7 cat% Y, a core diameter of 2 nm and an oxygen vacancy depleted space charge layer of about 5 nm width, we can simulate the doped YSZ ceramic by a brick model with a brick interior of bulk YSZ conductivity and walls that should represent the three regions of the space charge layer (space charge layer with very low oxygen vacancy concentration, a heavily doped layer of defective fluorite layer and a second space charge layer).

Influence Of Impurities

The studies by Badval et al.[1] of various 3YSZ ceramics illustrate that the bulk conductivity is rather independent of small variations in powder purity. Large variations in grain boundary conductivity, however, left room for speculations. Our extended studies on impurity segregation in YSZ ceramics suggest that the impurity level in a ceramic in general first decreases with firing temperature (loss of alkali and volatile impurities) and then increases during further firing with temperature and time due to the pick up of impurities by the ceramic from surrounding tube, holder and furnace materials. Thus we found that electrolyte in contact with silica-containing supports enriches in silica. For an overall average of 50ppm silicon or smaller and simultaneous presence of alumina, no glassy grain boundary films or second phase triple point pockets form. At high levels of silica, however, we observed the formation of silicate triple phase pockets and extended glassy films.

High temperature processing, firing of electrodes or contacts or long time impedance testing with electrodes and leads always implies the danger of introducing pollutants that are mobile in the grain boundaries. On a first view, platinum seems to be an inert material, ideal for contacting a clean electrolyte; however, ICP analysis of various batches and origins of Pt-paste showed that Pt-paste usually contains considerable amounts of silica, iron, calcium….. All these elements are present in small quantities in Pt-leads and -electrode contacts; they are mobile according to our diffusion studies and arrive with time during firing and electrochemical testing in the electrolyte and distribute in its grain boundaries or segregate to interfaces and surfaces[25]. Thus data obtained with Pt-paste contacts have to be carefully analyzed; especially an evolution of conductivity with time is often related to pollution of the system from outer sources.

We illustrated in our work an impurity induced change in grain boundary conductivity for the doped 3YSZ electrolyte. The Pd-Au-Pt electrical contacts contained 0.2 at% iron, 400ppm K and 150ppm Na and silicon as main impurities. Iron was mobile at the firing temperature of the contact, diffused into the electrolyte and segregated to the grain boundaries. As a result, a much lower grain boundary conductivity was measured compared to tests with (clean) Ag-Pd contacts, Figure 9c. The change in activation energy, Figure 10, indicates that other impurities besides iron have participated in the modification of the grain boundary chemistry..

CONCLUSIONS

In Mn, La and Fe-doped and undoped high purity 3YSZ and 8YSZ ceramics, spatially resolved TEM/EELS and impedance spectroscopy have been used to study the grain boundary segregation. All boundaries exhibit segregation. In undoped materials, yttrium segregates to the boundaries with enrichment factors of 2-3.5 in 3YSZ and 1.2-1.6 in 8YSZ. Significant enrichment of di- and trivalent ions occurs in doped materials. The enrichment of the cations is usually restricted to a grain boundary core of 1-2nm in width. Total concentrations of Mn^{2+}, Fe^{2+}, La^{3+} and Y^{3+} in the core can reach up to 40cat% in 3YSZ ceramics; in 8YSZ ceramics, those concentrations are by a factor of two to three lower. The reason for this difference may be the higher bulk solubility of the cations in 8YSZ and a lower coordination number in the cubic phase. Segregation factors were found to depend on the crystallographic structure of the boundary. O K ELNES features in the core are different from the bulk and suggest a highly defective fluorite-type bonding in the core.

Enrichment of the grain boundary core in lower valency dopants yields a negative grain boundary core charge. EELS of undoped 3YSZ revealed that the O K absorption edge height is increased in the region adjacent to the grain boundary core; it drops within 3-5nm distance to the bulk value. O K ELNES changes within this layer, exhibiting close to the core the features of tetragonal YSZ with very low dopant levels. The evolution in O K edge height and fine structure directly matches an oxygen vacancy depleted space charge layer and is considered as a direct observation of the latter.

Compared to 3YSZ ceramics, segregation is less pronounced in 8YSZ ceramics. O K does not reveal any significant modifications.

Inter- and intragranular resistance of doped and undoped ceramics has been deduced from impedance measurements. It is found that the oxygen ion bulk conductivity of YSZ ceramics remained unaffected by the dopants. The grain boundary resistance was significantly increased by di- and trivalent dopants. This was expected from the grain boundary chemistry. Both, formation of stable defect associates between oxygen vacancies and di- and trivalent cations in the grain boundary core and depletion in oxygen vacancies in the space charge layer contribute to the grain boundary resistance.

ACKNOWLEDGEMENTS

This work made use of the ultra-high vacuum scanning transmission electron microscopy and microscopy facilities of the Cornell Center for Materials Research (CCMR) with support from the National Science Foundation Materials Research and Engineering Centers (MRSEC) program (DMR-0079992). The authors want to especially thank M. Thomas, CCMR, for his support during the STEM measurements. They further want to thank K.Work, M.Carson and J.Brown, Corning Incorporated, for their technical contributions.

REFERENCES

[1] S.P.Badwal, F.T.Ciacchi; Ionics 6,1 (2005)

2 S.P.Badwal, Solid State Ionics 76, 67-80 (1995)

3 M.Ruehle, N.Claussen, A.Heuer; Advances in Ceramics Vol.12, p.352 (1984)

4 M.Backhaus-Ricoult, M.Badding, J.Brown, M.Carson, E.Sanford, Y.Thibault; Ceramic transactions 2AM-S8-55 (2004)

[5] High temperature solid oxide fuel cells, ed. Singhal, S.C., Kendall, K., Elsevier, Oxford (2004)

[6] X.Guo, J.Maier; J. Electrochem. Soc 148, E121-126 (2001)

[7] X.Guo, Solid State Ionics, 96, 247-254 (1997)

[8] J.Maier; Prog. Solid St. Chem. 23, 171-263 (1995)

[9] R.F.Egerton, Electron energy loss spectroscopy, Plenum Press, New York and London (1989)

[10] Electron Energy Loss Spectroscopy, R.F. Egerton, Plenum Press NY (1989)

[11] McDonald, Impedance spectroscopy (Chapter 4.1), John Wiley and sons, New York (1987)

[12] F.R.Chien, A.H.Heuer, Phil. Mag. A73, 681 (1996)

[13] H.Solmon, PhD thesis, CNRS- Labo. Phys. Mat. (1992)

[14] S.Weber et al. Applied Surface Science 203/204, 656 (2003)

[15] M.Kilo, M.Weller, G.Borchardt, B.Damson, S.Weber, S.Scherrer, Defect and Diffusion Forum 194-199, 1039 (2001)
[16] M.Kilo et al, J. Applied Physics 94, 7547 (2003)
[17] M.Kilo et al. Journal of Chemical Physics 121, 5482 (2004)
[18] C.T.Bak et al, J. Am. Ceram. Soc 85, 2244 (2000)
[19] K.Kovalsli, A.Bernasik, A.Sadovski, J. Eur. Ceram. Soc. 20, 951 and 2095 (2000)]
[20] M.Kilo et al. Journal of chemical physics 121, 5482 (2004)
[21] M.Backhaus-Ricoult, Y.Thibault; unpublished
[22] "Solid oxide fuel cells" by S.C. Singhal and K.KendallElsevier, Oxford (2003)
[23] I.M.Ross, W.M.Rainford, A.J.Scott, A.P.Brown, R.Brydson, D.W. McComb, ECERS 24, 2023-2029 (2004)
[24] E.C.Dickey, X.Fan, S.J.Penneycook, JACS 84, 1361-1368 (2001)
[25] M.Backhaus-Ricoult, M.-F.Trichet; Sol. St. Ionics, 150, 143 (2002)

Other Electronic Ceramic Applications

ELLECTRICALLY CONDUCTIVE MECHANISMS FOR Al_2O_3-C-TiCN CERAMICS

Hiroto Unno, Jun Sugawara and Toshio Mukai
New Materials Division, Nippon Steel Corporation, Tokyo 100-8071, Japan

ABSTRACT

The present authors have recently succeeded in developing the electrically conductive Al_2O_3 ceramics with a resistivity of $10^3\Omega cm$ by adding a small amount of carbon, TiCN and TiO_2. The effect of these additives on the electrical resistivity and the electrically conductive mechanism were investigated. This conductive Al_2O_3 ceramics was found to include an amorphous or multi-crystalline carbon and a TiCN phase. The solubility of titanium in Al_2O_3 was around 0.4 wt% TiO_2. Therefore, it is not probable that the high electrical conductivity is attributed to a conductive Al_2O_3 matrix. The carbon content had a significant effect on the resistivity, which showed a so-called percolation phenomenon thatexplains an abrupt resistivity change at a concentration threshold value of 0.35wt% carbon. We observed a conductive network structure by a unique SEM method. The particle boundary formed between spray-dried powder particles is considered to be enriched with carbon and contribute to a high electrical conductivity. TiCN was needed to maintain the conductive network; the grain growth of Al_2O_3 in lower TiCN contents leads to the destruction of ths network. On the other hand, the addition of TiO_2 in excess of 0.5 wt% indicated a negative influence on the electrical conductivity; TiO_2 promotes grain growth of Al_2O_3 and as a consequence the conductive network structure is destroyed.

INTRODUCTION

Al_2O_3 ceramics has been used in various structural and electrical applications, since it has excellent physical and mechanical properties. However, it is not suitable for applications requiring electrostatic dissipation, such as semiconductor preparation or fiber-handling equipments. In some cases of semiconductor preparation making use of electron beam, the electrical resistivity of materials is needed to be within the range of 10^1 to $10^4\Omega cm$.

Electrical conductivity can be enhanced by dispersion of conductive secondary phase particles over insulating Al_2O_3 matrix. Generally, TiC[1,2], TiN [3-5], metal particle[6,7] have been used as the dispersive phases, where the electrical resistivity of composites decreased to $10^0\Omega cm$ or less; this phenomenon could be explained by the percolation theory[8]. According to this theory, the electrical resistivity decreases abruptly around a certain threshold value of particle volume fraction, and the massive addition, generally over 15 volume percent, is required to obtain a high conductivity. Therefore, controlling the electrical resistivity of materials within the range of 10^1 to $10^4\Omega cm$ is rather difficult. Moreover, such materials sometimes exhibit poor characteristics in other properties, which result from inadequate dispersion process or the lack of sinterability caused by a large volume fraction of such particles.

Another method to enhance electrical conductivity is to add transition metal oxides, e.g., TiO_2[9], Cr_2O_3[10], to the Al_2O_3 matrix. Particularly, a small addition of TiO_2 has been widely studied[11-15], since it enhances densification and promotes grain growth during sintering process of Al_2O_3. The influence of TiO_2 on the electrical resistivity is closely re lated to solubility of titanium in Al_2O_3 matrix. Early studies have found that the solubility of titanium in Al_2O_3 exists between 0.15 and 1.8 mol%[11-15] with the electrical resistivity of usually $10^7\Omega$cm or more[9].

Thus, it has been difficult to obtain an electrically conductive Al_2O_3 ceramics having a resistivity from 10^1 to $10^4\Omega$cm up to now. However, the present authors have recently succeeded in developing these conductive ceramics by adding a small amount of carbon, TiCN and TiO_2 and firing in a reduction atmosphere as disclosed in the patent.[16] This research is aimed at evaluating the effect of these additives on the electrical resistivity and obtaining further insights of the mechanism for electrical conductivity.

EXPERIMENTAL PROCEDURE

As raw materials, Al_2O_3, TiCN and TiO_2 with purity higher than 99.5% having particle size of 0.5-1.0 μm are used. Carbon is added in a form of carbon black of 20nm in mean particle size. TiCN stands for Ti(C,N) with a NaCl structure having C to N ratio of 1:1.

Al_2O_3 powder and other additives were mixed together with an organic binder in an Al_2O_3 pot using a deionized water and zirconia media. After mixing, the slurry was dried using a spray dryer. The green compact of 100mm in diameter, 20mm in thicknesswas formed by cold isostatic pressing. The green compact was sintered at temperatures between 1650 and 1750°C for 4 h in an Ar-gas flow atmosphere. The heating rate is 3°C/min. During heating, an isothermal treatment at 300°C for 4h is allowed for burnout of the organic additives. In order to avoid the expansion of sample resulting from the carbothermal reaction of TiO_2 in the presence of carbon, the isothermal treatment at 1100°C for 5h was also carried out. Gas such as CO resulting from the carbothermal reaction of TiO_2 can be exhaled during the treatment.

The density of sintered samples was determined using the Archimedes method. The electrical resistivity of sintered samples was measured by the A.C. two-terminal method at a frequency of 10kHz by using a LCZ meter (HP-4276A, Hewlett Packard Co., USA). Some of these samples were also measured by the D.C. four-terminal method by using a galvanostat (1286, Solartron Co., UK). X-ray diffraction analyses (XRD; RAD-RB, Rigaku, Japan) were carried out to identify crystal structure and lattice parameter of each phase. Microstructures of the obtained samples were observed by scanning electron microscopy (SEM; S-2460N, Hitachi, Japan) operated at 20kV. The selected samples were observed in a unique condition without a conductive coating over the sample surface which allows us to identify the electrically conductive mechanism through the nature of absorption current. The samples were also examined by transmission electron microscopy (TEM; HF2000, Hitachi, Japan) operated at 200 kV. Samples for the TEM study were prepared by standard ion milling. Energy dispersive X-ray spectroscopy analyses of microstructure were performed in TEM using an EDS detector. Selected sample was also examined by atomic force microscopy (AFM; Nanoscope 3a, digital instruments, US) to study the electrically conductive mechanism.

RESULTS AND DISCUSSION

Microstructures

The present authors have obtained a high electrically conductive Al_2O_3 ceramics with a resistivity of $10^3\Omega$cm by adding 0.4wt% carbon, 3.0 wt% TiCN, 0.5 wt% TiO_2, 0.2wt% Y_2O_3, and 0.1wt% $MgCO_3$. Here, the sample having this composition is designated as "the standard sample". In order to study effects of the additives, the concentration of each additive (carbon, TiCN and TiO_2) was varied with the constant Y_2O_3 and $MgCO_3$.

The XRD pattern of the standard sample sintered at 1730°C indicated the presence of TiCN, ZrC, YAG, and α-Al_2O_3 phases with the absence of expected TiO_2 phase, as shown in figure 1.

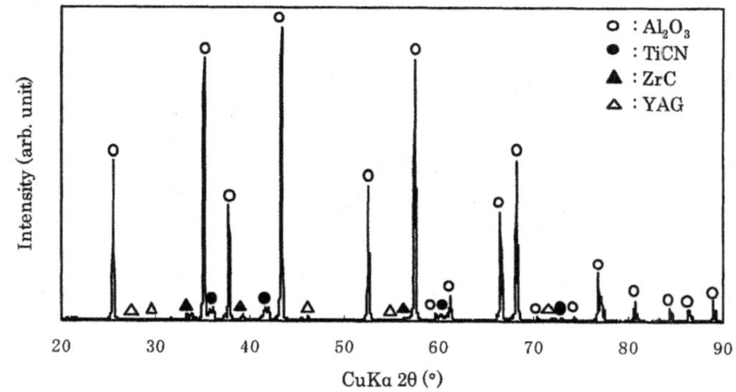

Fig. 1. XRD data for the standard sample of 0.5wt% TiO_2, 3.0wt% TiCN, 0.4wt% carbon, 0.2wt% Y_2O_3, 0.1wt% $MgCO_3$ balanced Al_2O_3 sintered at 1730°C.

The ZrC phase possibly came from zirconia media used at a powder mixing process; ZrO_2 converted into ZrC through a carbothermal reaction at high temperatures. It was found that the solubility of titanium in α-Al_2O_3 is small, since the X-ray patterns of α-Al_2O_3 showed no measurable peak shift. Excess TiO_2 have probably reacted with carbon and converted into a TiC phase. Figure 2 shows the X-ray diffraction peak intensities of TiCN (including TiC) as a function of TiO_2 content.

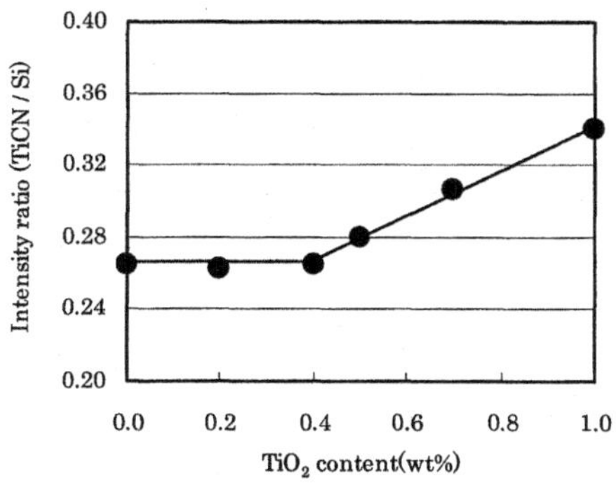

Fig. 2. Intensity of X-ray diffraction peaks of TiCN (200) relative to Si (111) in samples containing 3wt% TiCN and various amount of TiO_2 fired at 1730°C for 4h.

The intensity of TiCN increased beyond 0.4wt% TiO_2. Therefore the solid solution limit is considered to lie at around 0.4 wt% TiO_2. With this solubility, the electrical resistivity of Al_2O_3

matrix only decreases to $10^7\Omega$cm, as shown in a previous paper[9]. Therefore, it is not probable that the high electrical conductivity is attributed to a conductive Al_2O_3 matrix.

Figure 3 shows a SEM micrograph of the standard sample sintered at 1730°C with the conductive coating over the surface.

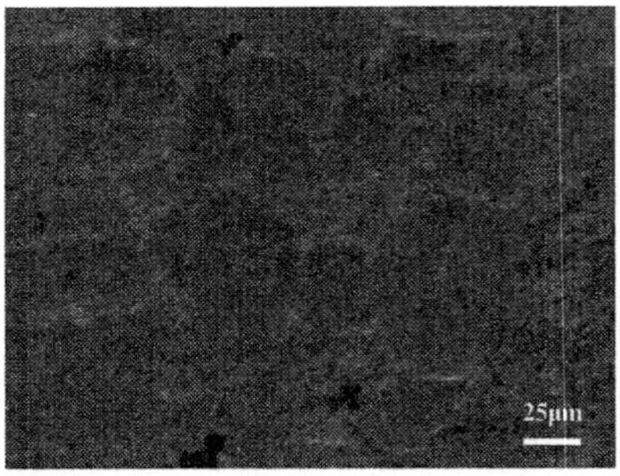

Fig. 3. SEM micropgaph for the standard sample.

The sample exhibits a 30 μm scale boundary which is thought to form between spray dried (SD) powder particles. A large amount of carbon was detected along the SD powder particle boundaries by EDS analysis. TiCN, and ZrC phases were uniformly dispersed in the Al_2O_3 matrix. Figure 4 shows a SEM micrograph of the above sample observed without the conductive coating over the surface at an optimum scanning speed operated at 20kV.

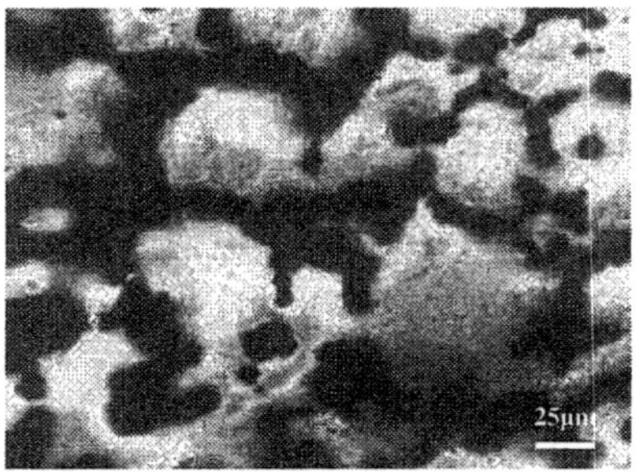

Fig. 4. SEM micrograph of the standard sample in a unique condition without a conductive coating over the surface.

The portions with higher resistivities are considered to emit more secondary electrons when excited by an electron beam, whereas the conductive parts appear to be black since electron can move rapidly in lower resistivity regions. The SD particle boundary concentrated with carbon, shown in black, indicates that it has a high electrical conductivity. In contrast, the central part of the SD particle, shown in white, indicates thatit has a higher resistivity compared to the boundary. The enrichment of carbon along the SD particle boundary can be developed by the nature of affinity between carbon and water during spray-drying process of slurry.

Effect of carbon

The effects of carbon content are represented in figure 5 (a) and figure 5 (b); the samples contain 3wt% TiCN and 0.5wt% TiO_2.

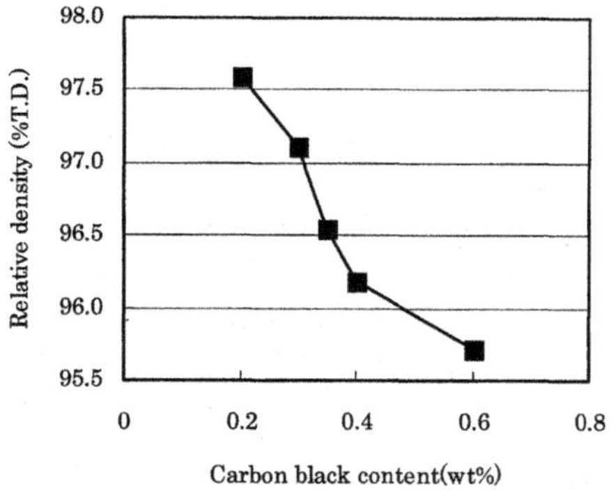

Fig. 5 (a). Relative density of samples containing various carbon black content fired at 1730°C for 4h.

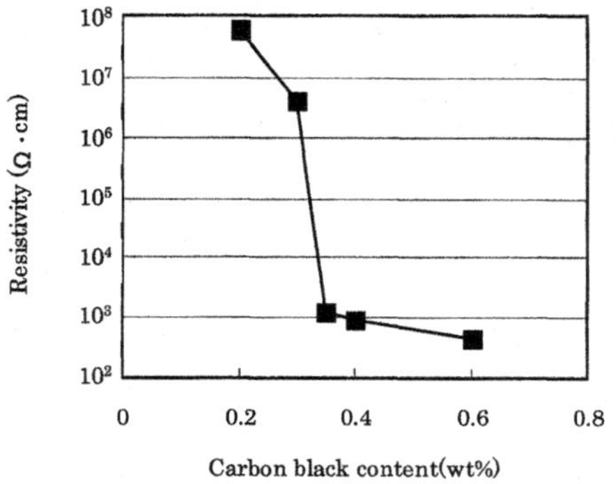

Fig. 5 (b). Resistivity of samples containing various carbon black content fired at 1730°C for 4h.

The density strongly decreased as the carbon content increased. This indicates that the carbon particle inhibits the densification of sintered bodies. The carbon content had a significant effect on the resistivity, which showed a so-called percolation phenomenon[8] that explains an abrupt resistivity change at a concentration threshold value of 0.35wt% carbon. Therefore, carbon is considered to be indispensable to attain the high electrical conductivity. It is probable that the high electrical conductivity is attained by the formation of a conductive network which is characterized by a carbon phase along the SD particle boundary.

Effect of TiCN

The effects of TiCN content are discussed. Figure 6 shows TEM micrographs of the standard sample sintered at 1730°C. The carbon phase was present in an amorphous or a multi-crystalline form in the grain boundary or the triple point of Al_2O_3 grains, as revealed by electron diffraction in figure 6 (b).

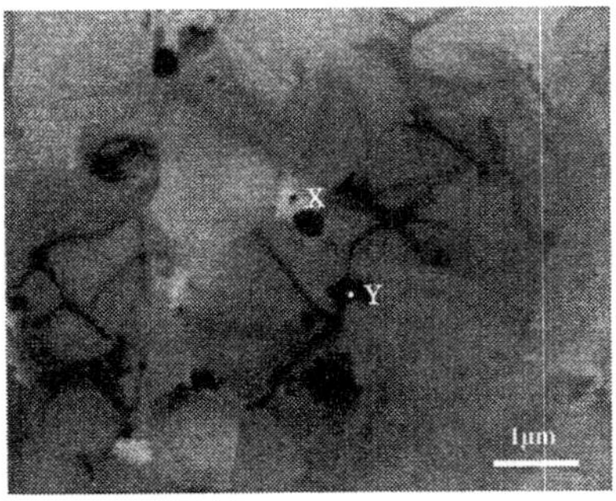

Fig. 6 (a). TEM micrograph of the standard sample sintered at 1730°C.

Fig. 6. (b) Electron diffraction patterns of points of X and Y.

TiCN and ZrC phases were also present inside the grain or along the grain boundary of Al_2O_3 grains in a round or faceted shape. The grain size of Al_2O_3 was 1 to 2μm. No Ti signal was detected within the Al_2O_3 grains. Therefore, the solubility of titanium in α-Al_2O_3 is small. This observation is in an agreement with the previous results of XRD analysis which indicates that Al_2O_3 has a solubility of Ti_2O_3 as small as 0.4wt%.

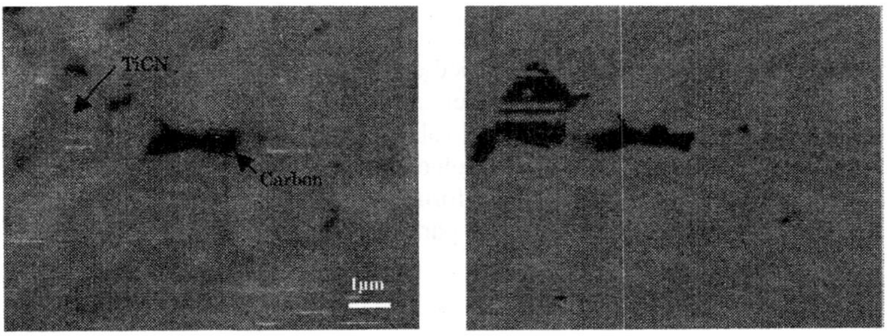

(a) (b)

Fig. 7. AFM analysis of the standard sample. (a) AFM micrograph, (b) Image of tunnel current.

Figure 7 (a) shows an AFM micrograph of the above sample. The image of tunnel current of the same field is shown in figure 7 (b). It reveals that the portions in accord with carbon and TiCN phases appear to be black corresponding to the high electrical conductivity. Therefore, carbon and TiCN phases are proved to contribute to the formation of the conductive network.

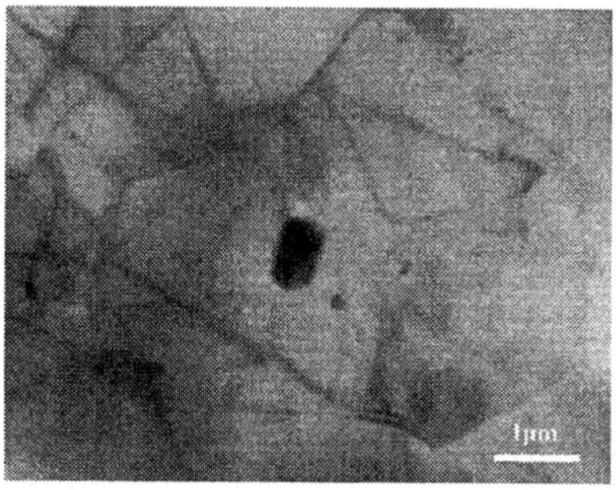

Fig. 8. TEM micrograph of a composition of 0.5wt% TiO_2, 0.4wt% carbon, Y_2O_3 0.2wt%, 0.1wt% $MgCO_3$ sintered at 1730°C.

Figure 8 shows a TEM micrograph of a high resistivity sample without TiCN-addition. The Al_2O_3 grain size is large compared to the case with TiCN-addition, which indicates the TiCN particle inhibits the grain growth. Although the carbon phase was found to be present along the Al_2O_3 grain boundary, the frequency was very few compared to the standard sample with TiCN-addition. The excess TiO_2 is considered to react with carbon and form TiCN.

The density is shown in Figure 9 (a) as a function of the TiCN content; the samples contain 0.4wt% carbon and 0.5wt% TiO_2 fired at various temperatures.

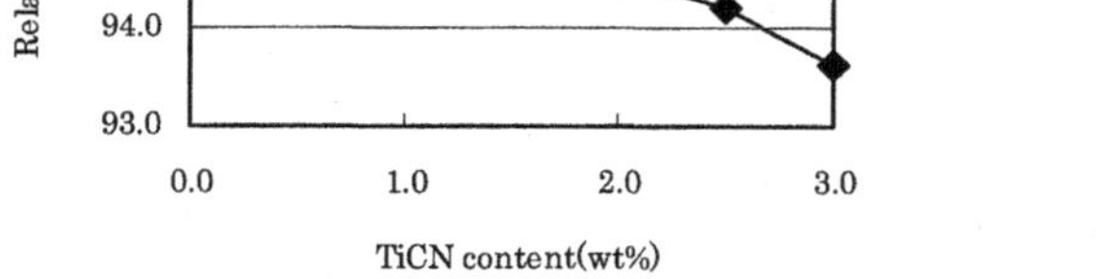

Fig. 9 (a). Relative density of samples containing various TiCN content fired at various temperatures for 4h.

The density slightly decreased as the TiCN content increased at every sintering temperature. The densification and grain growth of sintered bodies were inhibited by TiCN particles.

The electrical resistivity of the TiCN-containing samples is shown in figure 9 (b). In the case of high sintering temperatures, the resistivity was strongly dependent upon the TiCN content; it drastically decreased at around 1.5wt% TiCN, although, at low sintering temperature, the TiCN content had little or no effect on the resistivity. Consequently, when the densification of sintered bodies is not enough, in other words, the grain growth of Al_2O_3 does not proceed, the high electrical conductivity is attained.

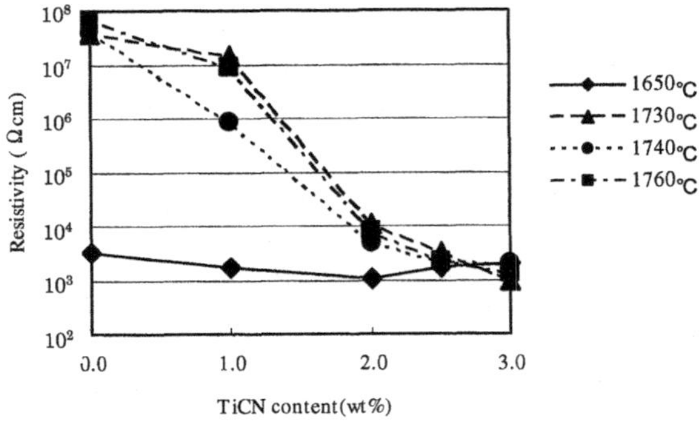

Fig. 9 (b). Resistivity of samples containing various TiCN content, 0.4wt% carbon and 0.5wt% TiO_2 fired at various temperatures for 4h.

The above result suggests that TiCN is necessary to maintain the conductive network containing carbon which is developed during spray-drying process of slurry. The grain growth of Al_2O_3 in lower TiCN contents leads to the destruction of this network.

Effect of TiO_2

The effects of TiO_2 on the density and electrical resistivity are discussed. The density is shown in Figure 10 (a) as a function of the TiO_2 content; the samples contain 3.0wt% TiCN and 0.4wt% carbon.

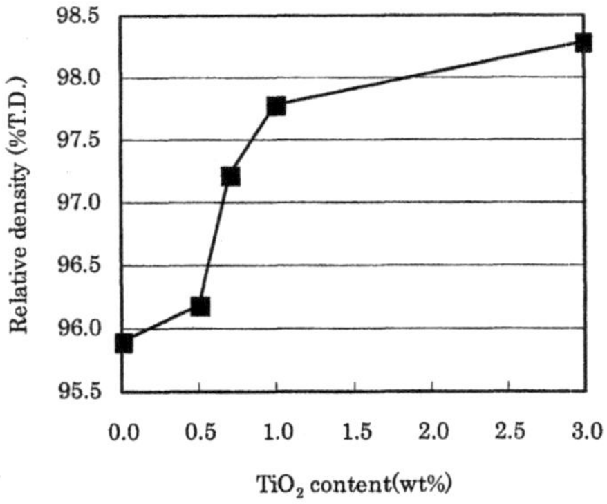

Fig. 10 (a). Relative density of samples containing various TiO_2 content fired at 1730°C for 4h.

The density increased as the TiO_2 content increased. This is because TiO_2 enhance densification via the grain growth during sintering, and the carbon phase, which inhibits densification of Al_2O_3, may be dissipated by carbothermal reaction with TiO_2. The electrical resistivity is shown in Figure 10 (b).

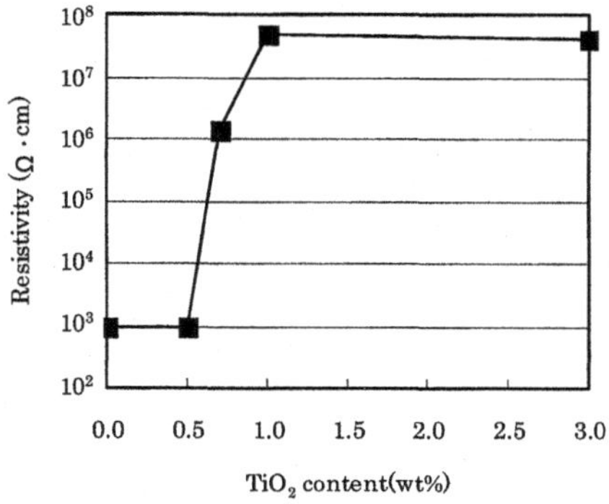

Fig. 10 (b). Resistivity of samples containing various TiO_2 content, 0.4wt% carbon and 3.0wt% TiCN fired at 1730°C for 4h.

The resistivity was strongly dependent on the TiO_2 content. However, unlike the case of TiCN and carbon, it drastically increased with increasing TiO_2 content. This is possibly because TiO_2 promotes grain growth of Al_2O_3, and as a consequence the conductive network is destroyed.

Electrically conductive mechanism

All the results indicate that the high electrical conductivity is attained by the formation of the conductive network containing carbon phase which has been developed during spray-drying process of slurry. Thus, it is essential that the grain growth of Al_2O_3 does not proceed, so that the densification of sintered bodies is insufficient.

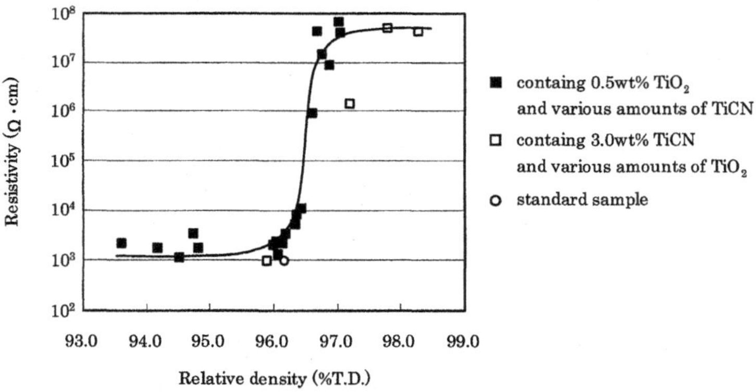

Fig. 11. The effect of the relative density on the electrical resistivity of samples containing 0.5wt% carbon and various amounts of TiCN and TiO_2.

Figure 11 shows the effect of the density on the electrical resistivity of the samples containing 0.4wt% carbon and various amounts of TiCN and TiO_2. All data points universally merge to a single continuous curve, which obviously indicates that the present considerations of electrically conductive mechanism are plausible. This also shows that the standard sample has a relatively high density compared to other samples having the same electrical conductivity.

In this study, the electrical conductivity is attained by a much less carbon-addition, compared to the general dispersive phases addition over 15 volume percent. Generally, to reduce the volume of dispersed phase, the approach to control the size of dispersed particles has been attempted. It is computationally and experimentally reported that the critical volume fraction changes with the ratio between a radius of dispersed phases R_m and matrix particles R_p[17, 18]. R. P. Kusy described the critical volume fraction V_c for the conductive phase in terms of the particle size ratio as

$$V_c = 100\left[1 + (\phi/4X_c)(R_p/R_m)\right]^{-1} \tag{1}$$

where Φ and X_c are the geometrical constant. Here, $\Phi = 1.27$ and $X_c = 0.42$ for many practical cases. In this study, the radius of dispersed phases R_m and that of matrix particles R_p are around 10nm for carbon and 15μm for SD particle, respectively. Therefore, the particle size ratio R_p/R_m is 1500 in the present case. Consequently, the critical volume fraction of the dispersed phase V_c derived from the above equation is estimated to be around 0.5 volume percent. This calculated value is in good agreement with the experimental value of this study; the electrical resistivity abruptly decreased at around 0.35wt% carbon.

CONCLUSION

Electrically conductive Al_2O_3 ceramics with a resistivity of $10^3\Omega cm$ was successfully fabricated by a small addition of carbon, TiCN and TiO_2 by firing in a reducing atmosphere. This conductive ceramics was found to include an amorphous or multi-crystalline carbon and a TiCN phase. The solubility limit of titanium in α-Al_2O_3 was around 0.4 wt% TiO_2. The carbon content had a significant effect on the resistivity, which showed a so-called percolation phenomenon that explains an abrupt resistivity change at a concentration threshold value of 0.35wt% carbon. We observed a conductive network structure by a unique SEM method. The particle boundary formed between spray dried powder particles is considered to be enriched with carbon and contribute to a high electrical conductivity. TiCN was needed to maintain the conductive network; the grain growth of Al_2O_3 in lower TiCN contents leads to the destruction of this network. On the other hand, the addition of TiO_2 in excess of 0.5 wt% indicated a negative influence on the electrical conductivity; TiO_2 promotes grain growth and as a consequence the conductive network structure is destroyed.

REFERENCES

[1]W. W. Gruss and K. M. Friederich, "Aluminum Oxide/Titanium Carbide Composite Cutting Tools"; pp. 48-62 in *Ceramic Cutting Tools.* Edited by E. D. Whitney. Noyes Publications, Park Ride, NJ, 1994.

[2]S. J. Burden, "Comparison of Hot-isostatically Pressed and Uniaxially Hot-Pressed Alumina-Titanium Carbide Cutting Tools," *Am. Ceram. Soc. Bull.*, **67** 1003-05 (1988).

[3]Y. G. Gogotsi and F. Porz, "Mechanical Properties and Oxidation Behavior of Al_2O_3-TiN Composites," *J. Am. Ceram. Soc.*, **78** 2251-59 (1992).

[4]Z. S. Rak and J. Czechowski, "Manufacture and Properties of Al_2O_3-TiN Particulate Composites," *J. Eur. Ceram. Soc.*, **18** 373-80 (1998).

[5]K. Tsugeki, T. Kato, Y. Koyanagi, K. Kusakabe, and K. Morooka, "Electroconductivity of Sintered Bodies of α-Al_2O_3-TiN Composite Prepared by CVD Reaction in a Fluidized Bed," *J. Mater. Sic.*, **28** 3168-72 (1993).

[6]R. Gunther, T. Klassen, B. Dickau, F. Gartner, A. Bartels and R. Bormann, "Advanced Alumina Composites Reinforced with Titanium-Based Alloys," *J. Am. Ceram. Soc.*, **84** 1509-13 (2001).

[7]N. J. Welham, P. E. Willis and T. Kerr, "Mechanochemical Formation of Metal-Ceramic Composites," *J. Am. Ceram. Soc.*, **83** 33-40 (2000).

[8]D. S. McLachlan, M. Blaszkiewicz, and. E. Newnham, "Electrical Resistivity of Composites," *J. Am. Ceram. Soc.*, **73** 2187-203 (1990).

[9]T. Watanabe, T. Aoshima, T. Kitabayashi and C. Nakayama, "Resistivity and Microstructure of Alumina Ceramics Added with TiO_2 Fired in Reducing Atmosphere," *J. Ceram. Soc. Japan,* **101** 1107-14 (1993).

[10]T. Watanabe and C. Nakayama, "Sintering of Alumina Green Sheet with TiO_2 and Cr_2O_3 in Reducing Atmosphere," *J. Ceram. Soc. Japan,* **101** 556-63 (1993).

[11]S. K. Roy and B. L. Coble, "Solubilities of Magnesia, Titania, and Magnesium Titanate in Aluminum Oxide," *J. Am. Ceram. Soc.*, **51** 3-6 (1968).

[12]R. D. Bagley, I. B. Cutler and D. L. Johnson, "Effect of TiO_2 on Initial Sintering of Al_2O_3," *J. Am. Ceram. Soc.*, **53** 136-41 (1970).

[13]W. D. Mckee Jr. and E. Aleshin, "Aluminum Oxide-Titanium Oxide Solid Solution," *J. Am. Ceram. Soc.*, **47** 54-58 (1964).

[14]E. R. Winkler, J. F. Sarver and I. B. Culter, "Solid Solution of Titanium Dioxide in Aluminum Oxide," *J. Am. Ceram. Soc.*, **49** 634-37 (1966).

[15]N. G. Schrewelius, "Synthetic Star Rubies and Star Sapphires," *J. Am. Ceram. Soc.*, **31** 170-75 (1948).

[16]H. Unno, N. Kosugi and J. Sugawara, *Japanese Patent Application* No. 2004-18296.

[17]A. Malliaris and D.T. Tuner, "Influence of Particle Size on the Electrical Resistivity of Compacted Mixtures of Polymeric and Metallic Powders," *J. Appl. Phy.*, **42** 614-18 (1971).

[18]R. P. Kusy, "Influence of Particle Size Ratio on the Continuity of Aggregates," *J. Appl. Phy.*, **48** 5301-05 (1977).

DIELECTRIC PROPERTIES OF HIGH-K LTCC MATERIALS

Jean-Pierre Ganne, Michel Pate, Olivier Durand, Claude Grattepain
Thales Research and Technology
Domaine de Corbeville
91404 Orsay, France

ABSTRACT

LTCC substrates are widely used for RF and microwave applications. In order to integrate capacitors, high permittivity materials compatible with the established LTCC technology are needed.

We prepared $Bi_2Zn_{2/3}Nb_{4/3}O_7$ (orthorhombic BZN) and $Bi_{1.5}ZnNb_{1.5}O_7$ (cubic BZN) powders by the classical powder-mixing and calcining method. These powders served as the basis to produce composite ceramics with different percentages of orthorhombic and cubic phases. The resulting materials were characterized with XRD, SEM, EDS, and dielectric measurements: temperature dependence of relative permittivity at 10 MHz, dielectric constant and losses at 3.5 GHz.

Orthorhombic BZN presents a high dielectric constant (k = 80) with a positive temperature coefficient (220 ppm/°C) and low losses at 3.5 GHz (about 10^{-3}). Cubic BZN presents a higher dielectric constant (k = 150), with a negative temperature coefficient (-500 ppm/°C) and rather high losses at 3.5 GHz (about 4×10^{-2}).

The composite materials made by mixing different amounts of cubic powder in orthorhombic powder present intermediate dielectric properties, but their variation versus the percentage of cubic powder is not regular, showing rapid changes between 20 % and 40%. XRD and EDS analyses show that the percentages of the two phases in the composite materials is affected by sintering. Plotting the dielectric properties versus the percentage of cubic phase determined by quantitative XRD analysis leads to a surprising result, a jump at 30% of cubic phase. This anomaly cannot be explained using the classical models for dielectric properties of mixtures. A simple model based on the percolation of the cubic phase fits well with the experimental data.

INTRODUCTION

Bi-Zn-Nb pyrochlores are being studied in many laboratories mainly as potential materials for integration of stable capacitors into LTCC substrates[1-4]. Numerous other studies are devoted to BZN: some are related to thin films[5,6] , aims to investigate the phase diagram[7] or to lower the sintering temperature[8-9] but we do not know articles dedicated to microwave characterisation of the material.

The object of our study was to optimize the thermal stability of the material while keeping the relatively high dielectric constant and low dielectric losses. Two phases draw the attention: $Bi_2Zn_{2/3}Nb_{4/3}O_7$ (orthorhombic BZN) and $Bi_{1.5}ZnNb_{1.5}O_7$ (cubic BZN). Orthorhombic BZN (O-BZN) presents a high dielectric constant (k = 80) with a positive temperature coefficient (220 ppm/°C), while Cubic BZN (C-BZN) presents a higher dielectric constant (k = 150), with a negative temperature coefficient (-500 ppm/°C). By mixing O-BZN and C-BZN powders it is possible to make composite ceramics with intermediate dielectric properties, depending of the percentages x and 100-x of C-BZN and O-BZN in the composite. The percentages x about 30 % to 40 % are of special interest as the expected temperature coefficient is near zero.

We prepared composites with x = 10 %, 20%, 30%, 40 % as initial percentages, and single-phase ceramics O-BZN (x = 0 %) and C-BZN (x = 100 %), and characterised them for their chemistry and phases composition and microstructure with XRD, SEM, EDS, and we measured their dielectric properties.

EXPERIMENTAL

Material Synthesis:

Powders of O-BZN and C-BZN were prepared following the standard ceramic process. Raw materials (Bi_2O_3 Aldrich n° 22389-1, ZnO Merck n°1-08846, Nb_2O Alfa Aesar n°11366) are weighed in stoichiometric proportions, mixed in suspension in water with a Turbula mill with polyethylene balls. After evaporating water, the powder mixture is calcined for 2 hours, in order to react the powders. After calcining, the powder is milled with a Labstar recirculation mill for 50 minutes and calcined a second time in the same condition as the first calcination.

Calcining temperatures were determined by thermal analyses and XRD for phase purity: they are respectively 800 °C and 950 °C for O-BZN and C-BZN, which is in agreement with the results of Xiao-Li and al[3] showing that the orthorhombic phase is more stable under 800°C while the formation of cubic phase is favoured above 950°C.

In order to synthesize the composites, the powders of O-BZN and C-BZN were weighed in molar percentages 100-x and x (x = 0, 10, 20, 30, 40 and 100) and milled with zirconia balls in water in a Labstar recirculation mill for 50 minutes. After drying and sieving, pellets were pressed and sintered. The optimal sintering temperatures are different for pure O-BZN (875 °C for 2 hours), and composites or pure C-BZN (975 °C for 2 hours). The measured densities are respectively 7.67 (which is 96.6 % of the theoretical density), and 6.9 (97 % of the theoretical density) for O-BZN and C-BZN.

Dielectric Characterisation:

Two kinds of samples were used: 12 mm diameter, 0.5 mm thick disks metallized with silver electrodes for capacitor measurements at low frequencies (100 Hz – 100 kHz) , and non-metallized cylinder for microwave measurements at 3.5 GHz.

The capacitor measurements at 100 kHz were made with an impedance-meter (HP 4194) and a N2 liquid cryostat and a microfurnace in order to determine dielectric constant, losses and temperature coefficient of capacitance TCC.

The microwave measurements were made at ambient temperature.

For samples presenting low losses (Tan $\delta < 1$ %) we can use dielectric resonators techniques. When dielectric losses are higher, we use a coaxial transmission technique.

Microstructure and physico-chemical characterisation:

The X-ray analyses are performed using a diffractometer equipped with a 120° CPS detector with a front germanium monochromator. EDS and micrographic observations are done with a Topcon dual stage scanning electron microscope.

RESULTS

Dielectric properties versus x:

Both TCC (fig1) and microwave dielectric properties (fig 2) showed an irregular dependence versus x, with a very slow variation for x < 20 %, a rapid change between 20 and 40 %, and a slower variation between 40 and 100 %. Here "x" is the nominal molar percentage of C-BZN powder in the composite. We will show later that it may be different of the true molar percentage of C-BZN "X" in the composite ceramic materials, because the sintering treatments change the proportions between the phases.

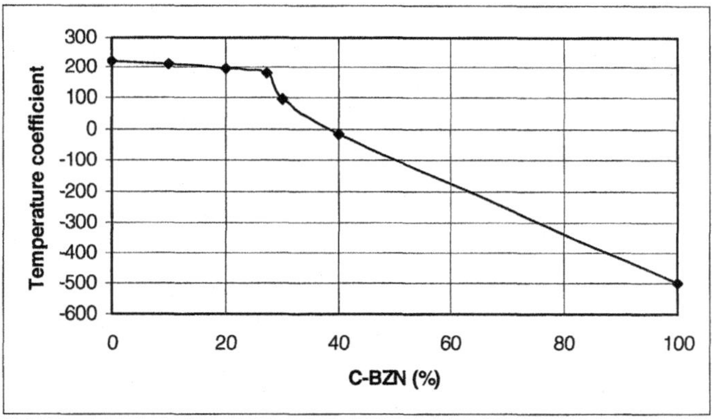

Figure 1 : Temperature Coefficient of Capacitance of BZN composites versus x, percentage of C-BZN powder in the composite.

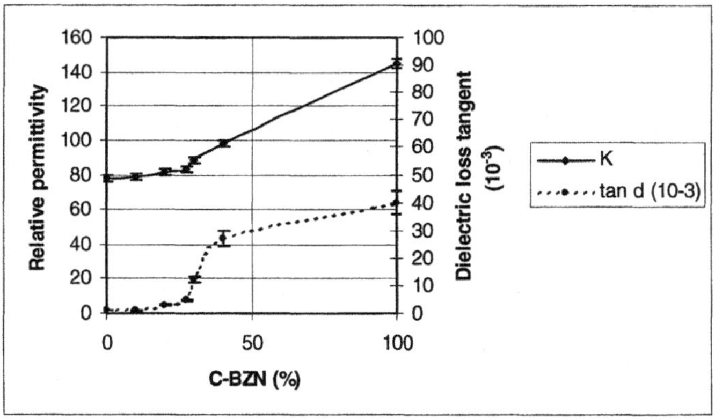

Figure 2 : Relative permittivity and dielectric loss tangent at 3.5 GHz of BZN composites versus x, percentage of C-BZN powder in the composite.

A first qualitative series of XRD spectra was carried out on composites with different percentages. The main diffraction lines of C-BZN were absent for composites with the lower percentages x, up to 30 %, and then they increased rapidly between 30 and 40 %, as shown in Fig 3.

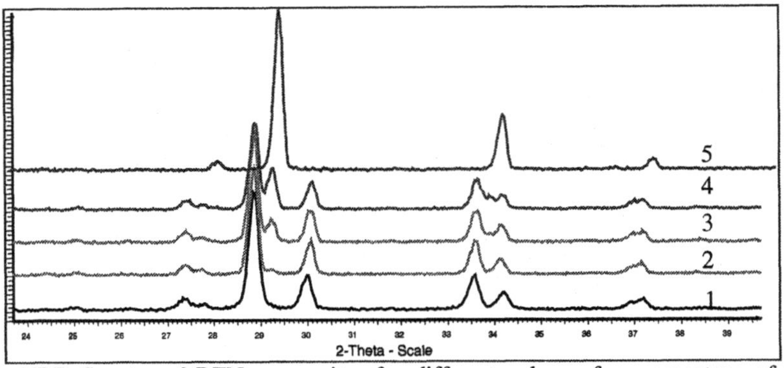

Figure 3: XRD Spectra of BZN composites for different values of x, percentage of C-BZN powder in the composite (1: x = 0, 2: x = 20%, 3 :x = 30%, 4: x = 40%, 5: x = 100%).

This seemed to be in good agreement with the results of dielectric measurements, but in contradiction with the results[3] of literature: the relatively high sintering temperatures used for composites should favour the formation of more cubic phase than initially present in the powder mixtures.

We decided to make more extensive analyses of the composites: XRD with a quantitative determination of the proportion of phases, SEM and EDS in order to study the repartition of phases and their relation with local composition.

Determination of phase proportions by XRD:

The real amounts of C-BZN phase in the composite were determined by XRD analysis as the ratio of the total areas of all the diffraction peaks of this phase versus the sum of the areas of all the peaks of the two phases. In order to correct the structure factor, these results were corrected using a calibrating factor obtained with the measurement in the same conditions of the ratio of C-BZN in the mixtures of the two powders. The ratios X obtained with this method differ from the expected ones.

Table 1 shows the results of this procedure, with the true molar percentage X of C-BZN in the final ceramic material as a function of the nominal percentage x introduced in the mixed powders before pressing and sintering.

x (initial powders)	0	10	20	27.5	30	40	100
X (sintered ceramic)	0	16	28	27	28	29	100

Table 1: Molar percentage X of C-BZN in ceramics versus initial percentage x in powder mixtures

One observes a higher C-BZN content in the composites with a starting percentage x below 30% and a lower C-BZN content above this percentage.

The corrected evolutions of the TCC, permittivity and dielectric losses of BZN composites versus X are shown figures 4 and 5.

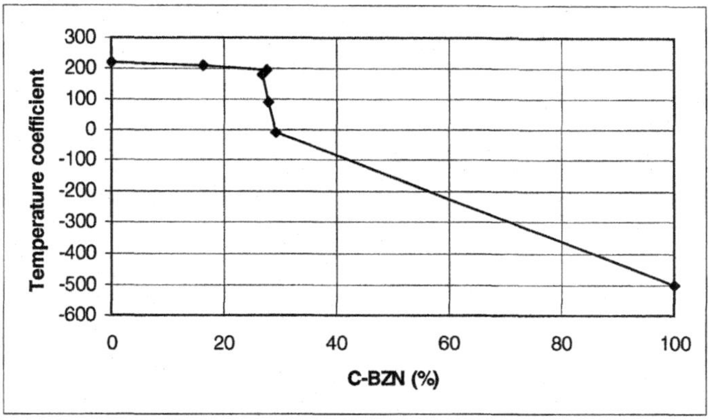

Figure 4 : Temperature Coefficient of Capacitance of BZN composites versus measured percentage X of C-BZN.

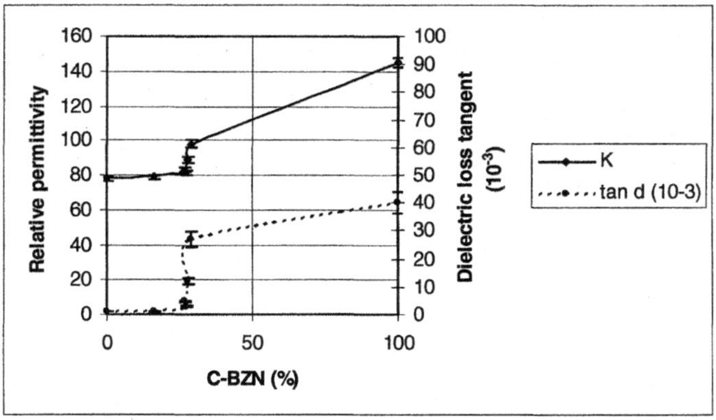

Figure 5 : Relative permittivity and dielectric loss tangent at 3.5 GHz of BZN composites versus measured percentage X of C-BZN.

EDS analysis

EDS analyses were performed in order to identify the two phases in the composites. SEM micrographs were obtained for x = 10% and 40% composites. An image of the x = 40 % material is shown figure 6. One can see dark and light grains. No such contrast can be seen on the x = 10 % composite. Comparison between the theoretical composition of the elements, calculated taking into account the initial ratios of the two phases written as : $Bi_b\ Zn_z\ Nb_b\ O_7$ and the measured compositions is done in table 2. For the x = 40% composite the composition of the two different grain families are reported.

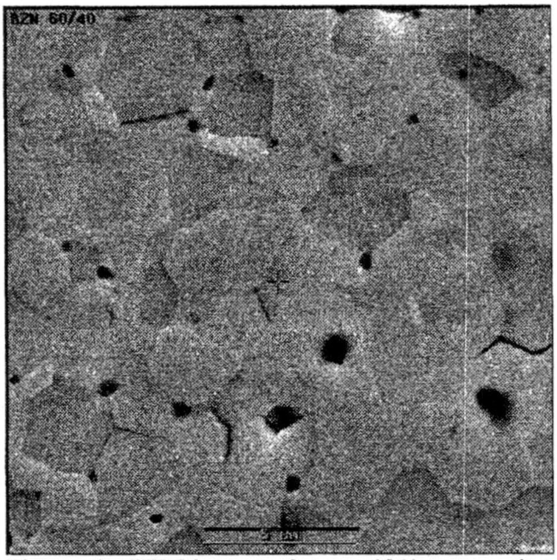

Figure 6 : SEM image of a x = 40% composite.

	Bi(b)		Zn(z)		Nb(n)	
x	theoretical	measured	theoretical	measured	theoretical	measured
0.1	1.95	1.89 ± 0.01	0.7	0.66± 0.01	1.35	1.4± 0.01
0.2	1.9	1.86± 0.01	0.73	0.72± 0.01	1.37	1.4± 0.01
0.275	1.86	1.82± 0.01	0.76	0.74± 0.01	1.38	1.41± 0.01
0.3	1.85	1.79± 0.01	0.77	0.77± 0.01	1.38	1.42± 0.01
0.4 dark grains	1.8	1.75± 0.01	0.8	0.73± 0.01	1.4	1.46± 0.01
0.4 light grains	1.8	1.65± 0.01	0.8	0.84± 0.01	1.4	1.47± 0.01

Table 2 : Comparison between theoretical and EDS measured composition of the composites.

These results shows a loss of bismuth and zinc in all the composites. During the sintering, the compositions of the two phases tend to homogenise. For the x = 40% composite one can identify a cubic rich phase (the light grains present a higher content in Zn and lower content in Bi).

DISCUSSION

The presence of a well-defined step at X = 30 % in the curves, with a transition from composites (X < 30 %) whose properties are close those of the pure O-phase, to composites (X > 30 %) whose properties seem to be dominated by those of the C-phase, suggest the possibility of a percolation-like phenomenon of the cubic phase inside the orthorhombic phase.

In usual percolation there is an infinite contrast between the properties (the conductivity) of both phases. Here the analogue of the conductivity is the dielectric constant, but the contrast is only a factor 2 between C-BZN (K= 140) and O-BZN (K=80). However the microwave admittance of the C-phase is indeed two times higher than that of the O-phase. The consequence is that we can imagine two cases:

1) for X < 30 %, grains of C-phase with the higher admittance are isolated in the matrix of O-phase, with a lower admittance. The equivalent electrical model for the two phases is two capacitors in series.
2) For X > 30 %, there is enough C-phase, so that we can always find paths of C-Phase grains interconnected, like in a percolation phenomenon. The equivalent electrical model for the two phases is two capacitors in parallel.

Simple equivalent circuit calculations were made for X = 30 % and compared with experimental data for X approaching 30 % from lower or upper values. Table 3 shows that there is a qualitative agreement between calculations and experimental data, which is satisfactory taking into account the oversimplification of the model.

F=3.5 GHz	Dielectric Properties of Pure Phases		Equivalent Circuit Calculations (X = 30 %)		Experimental Data		
	O-BZN	C-BZN	Series Model	Parallel Model	X <= 30 % (x = 27.5)	X = 30 % (x = 30)	X >= 30 % (x = 40)
K	80	140	92	98	83	89	98
Tan δ	1×10^{-3}	4×10^{-2}	8.7×10^{-3}	17.7×10^{-3}	5×10^{-3}	12×10^{-3}	27×10^{-3}

Table 3: comparison between the measured dielectric properties of the composites around X = 30 % and the calculations of a simple model (see text).

CONCLUSIONS

BZN composite ceramics were synthesised and characterised for their microstructure, composition and percentages of phases and elements, and dielectric properties. The compositions of ceramics determined by EDS are slightly different of the nominal composition, showing departure of zinc and bismuth. It was confirmed that it is possible to produce ceramics with zero TCC, but their dielectric losses are high at microwave frequencies.

The variation of dielectric properties with the percentage X of cubic BZN was studied. All the dielectric characteristics (K, Tan δ, TCC) show a rapid variation near X = 30 %. For X < 30 % the dielectric properties are dominated by the properties of the orthorhombic phase, while they are determined by the properties of the cubic phase for X > 30 %. A simple model is proposed in order to explain this behaviour. It is in qualitative agreement with the experimental data.

REFERENCES

[1]Man F.Yan, Hung C.Ling, Warren W. Rhodes, "Low Firing, Temperature-stable Dielectric Compositions Based on Bismuth Nickel Zinc Niobates", *J. Am. Ceram. Soc.*, **73**, 1106-107 (1990).

[2]Donhang Liu, Yi Liu, Shui-Q. Huang, Xi Yao, "Phase Structure and Dielectric Properties of Bi_2O_3-ZnO-Nb_2O_5-Based Dielectric Ceramics", *J. Am. Ceram. Soc.*, **76**, 2129-32 (1993).

[3]Xiao-Li Wang, Hong Wang, Xi Yao, "Structures, Phase Transformations, and Dielectrics Properties of Pyrochlores Containing Bismuth", *J. Am. Ceram. Soc.*, **80**, 2745-48 (1997).

[4]Matjaz Valant and Peter K. Davies, "Crystal Chemistry And Dielectric Properties of Chemically Substituted $(Bi_{1.5}Zn_{1.0}Nb_{1.5})O_7$ and $Bi_2(Zn_{2/3}Nb_{4/3})O_7$ Pyrochlores", *J. Am. Ceram. Soc.*, **83**, 147-53 (2000).

[5]Dong Hyuk Back, Yoon Seop Lee, Young Pyo Hong, Joong Ho Moon, Kyung Hyun Ko, "Tunability of Bi-rich cubic pyrochlore thin films by reactive sputtering", *Materials Research Society Symposium Proceedings*, **783**, 115-120 (2004).

[6]Hsiu-Fung Cheng, Yi-Chun Chen, Hsiang-Lin Liu, Luu-Gen Hwa, I-Nan Lin, "Correlation of microwave dielectric properties and crystallinity for pulsed laser deposited $Bi_2(Zn_{1/3}Nb_{2/3})_2O_7$ thin films", *Journal of the European Ceramic Society*, **24**(6), 1791-1794, (2004).

[7]Hong Wang, Xiao Li Wang, Xi Yao, "Phase equilibrium in Bi_2O_3-ZnO-Nb_2O_5 system", Ferroelectrics, **195**(1-4), 19-22, (1997).

[8]Kyeong Ae Yee, Kyoung Ran Han, Hyo Tae Kim, "The effect of V_2O_5 on the sinterability and physical properties of Bi_2O_3-NiO-Nb_2O_5 and Bi_2O_3-ZnO-Nb_2O_5 temperature-stable dielectrics", *Journal of Materials Science*, **34**(19), 4699-4704, (1999).

[9]Hong Wang, Huiling Du, Zhen Peng, Meiling Zhang, Xi Yao, "Improvement of sintering and dielectric properties on Bi_2O_3-ZnO-Nb_2O_5 pyrochlore ceramics by V_2O_5 substitution", *Ceramics International*, **30** (7), 1225-1229, (2004).

MONOLITHIC INTEGRATION OF NONLINEAR $BA_{1-X}SR_XTIO_3$ THIN FILMS WITH AFFORDABLE SILICON SUBSTRATES FOR FREQUENCY AGILE MICROWAVE DEVICE APPLICATIONS

M.W. Cole, W.D. Nothwang
U. S. Army Research Laboratory, Weapons and Materials Research Directorate, Aberdeen Proving Ground, Maryland, Email: mcole@arl.army.mil

R. G. Geyer
National Institute of Standards and Technology, RF Technology Division, Boulder, Colorado

ABSTRACT

This study demonstrates that large area affordable silicon substrates can be utilized in phase shifter devices if a "*microwave friendly*" passive buffer layer is sandwiched between the silicon substrate and the low loss active $Ba_{1-x}Sr_xTiO_3$ (BST) based thin film. In this heterostructure configuration there will be negligible attenuation of the microwave energy by the substrate. In our approach, metalorganic solution deposition (MOSD) fabricated Ta_2O_5 was employed as a buffer layer film to promote the integration of BST with affordable large area silicon substrates. In addition, the buffer layer material was designed to promote a tensile stress in the active thin film, such that the tunability of the BST-based coplanar device was enhanced to levels comparable to that of a varactor device design. The optimized Ta_2O_5 based thin film possessed excellent material properties, namely, an enhanced dielectric constant (ε_r =45.6), low dielectric loss (tan δ=0.006), high film resistivity ($\rho=10^{12}$ Ω-cm at E=1 MV/cm), excellent temperature stability (TCK of 52 ppm/°C), and excellent bias stability of capacitance (~1.41% at 1 MV/cm). The integration of the BST/Ta_2O_5 bilayer heterostructure on silicon substrates was demonstrated via a dual anneal process protocol. The defect free, structurally abrupt BST/Ta_2O_5 and Ta_2O_5-silicon substrate interfaces; combined with the smooth, uniform and defect free surface morphology of the dual anneal fabricated heterostructure suggests this process protocol to be an excellent method for achieving successful monolithic integration of BST with affordable silicon substrates in a coplanar device design.

INTRODUCTION

There is a demand from both the military and industrial communications sectors to produce integrated microwave phase shifters for mobile on-the-move antennas which are small size, low power consumption, and affordable. There is simultaneously a requirement for frequency agile tunable phase shifter devices to possess enhanced performance, i.e. low loss, wide tunablity and long term reliability. In this work we demonstrate a design and process science technology that satisfies both demands, by utilizing acceptor doped $Ba_{1-x}Sr_xTiO_3$ (BST) thin films which are monolithically grown on low cost Si substrates. We have previously demonstrated excellent microwave properties for Mg acceptor doped BST thin films on ceramic MgO substrates.[1,2] These films offer high tunability, low loss, and low tuning voltages from 100 kHz to the mid-microwave frequency range. In this work we demonstrate the successful integration of Mg acceptor doped thin films with low lost silicon substrates.

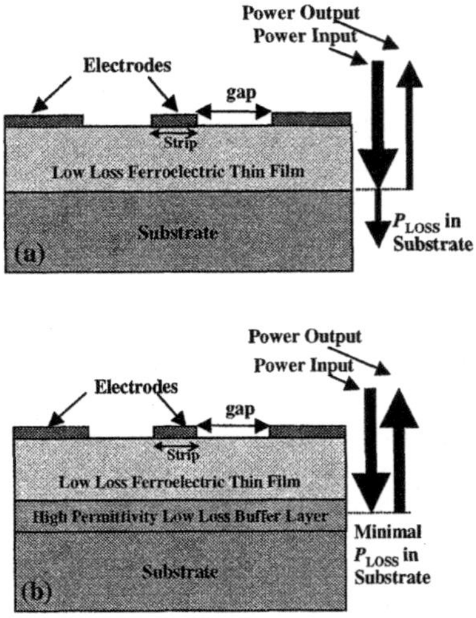

Figure 1. Coplanar device design; (a) tunable low loss active thin film deposited on lossy substrate and (b) tunable low loss active thin film integrated with the same lossy substrate utilizing a "*microwave friendly*" buffer layer.

Mobile antenna performance specifications (e.g., low insertion loss), requires the phase shifter substrate to be "*microwave friendly*", that is, one which does not attenuate the microwave energy of the actual device material and structure. Additionally, the substrate must be large area ($\geq$ 4 inch diameter), low cost, thermally stable, chemically resistant, industry standard, structurally suitable with good mechanical integrity, and must not perturb the performance of the active layer(s) constituents of the phase shifter device. Si wafers are the primary candidate for such a carrier substrate. The interest in Si as a carrier substrate is driven by the affordability and large size availability of Si wafers and the widespread industrial/commercial use of Si based processing technology. However, for microwave device applications, such integration precludes the use of bare Si substrates since Si is not a "*microwave friendly*" material, i.e., at microwave frequencies Si becomes less resistive, and consequently is too lossy for use in microwave devices. Therefore, Si is not the semiconductor of choice for microwave components. Figure 1a depicts a standard coplanar microwave phase shifter design whereby a microwave friendly, low loss tunable paraelectric active thin film is deposited on a "non-microwave friendly" substrate such as Si. This configuration is performance limiting since there is considerable attenuation of the microwave energy by the substrate, i.e., significant power loss in substrate. However, in theory, silicon substrates can be utilized if "*microwave friendly*" passive buffer layers can be sandwiched between the Si substrate and the active paraelectric thin film. Figure 1b is representative of the configuration where a "*microwave friendly*" passive (non-tunable) buffer layer is sandwiched between the Si substrate and the low loss active (tunable) thin film. In this case there is negligible attenuation of the microwave energy by the substrate.

In this work we demonstrate that silicon substrates can be utilized in phase shifter devices if a "*microwave friendly*" passive buffer layer can be sandwiched between the Si substrate and the low loss active paraelectric BST based thin film. In this heterostructure configuration there is negligible attenuation of the microwave energy by the substrate. In addition, if the buffer layer is designed to promote a tensile stress in the active thin film, the tunability of the device will be enhanced to levels comparable to that of varactor device design.

EXPERIMENTAL

The Ta_2O_5 buffer layer thin films were fabricated via the metalorganic solution deposition (MOSD) technique. Tantalum ethoxide was used as the precursor to form the film and acetic acid and 2-methoxyethanol were used as solvents and for viscosity adjustment. The precursor films were spin coated onto Pt-coated silicon and bare n^+-silicon substrates. Prior to film deposition, the silicon substrates were cleaned by spin etching.[3] This cleaning allowed for the removal of the native silicon oxide and hydrogen termination of the Si dangling bonds. Particulates were removed from the solution by filtering through 0.2 μm syringe filters. Subsequent to coating, the films were pyrolyzed at 350 °C for 10 min. in order to evaporate solvents and organic addenda to form an inorganic amorphous film. The spin coat pyrolyzation process was repeated until a nominal film thickness of ~100 nm was achieved. Optimized film crystallinity was achieved via post-deposition annealing in an oxygen ambience at 750 °C for 1 hour.

The coplanar device heterostructure configuration, depicted in Figure. 1b, was achieved by depositing a 200 nm 5 mol-% Mg doped BST paraelectric film, over the optimized Ta_2O_5 buffer layer. The experimental details of details of the Mg doped BST film fabrication have presented elsewhere.[2] The heterostructure fabrication method was optimized via a dual anneal process science protocol. The dual anneal process protocol consisted of depositing a 100 nm layer of Ta_2O_5 on a Si substrate, pyrolyzing (350 °C), and annealing at 750 °C for 60 minutes. Subsequent to the Ta_2O_5 film anneal a 200 nm of Mg doped BST film was deposited and pryrolyzed at 350 °C. The composite heterostructure was then annealed for 60 min. at 750 °C in an oxygen ambiance.

The dispersion of dielectric response (capacitance and loss tangent) of the of the Ta_2O_5 thin films was measured as a function of frequency (1 kHz to 1 MHz), temperature (25-125 °C) and applied bias (-15 to +15 V) in the metal-insulator-metal (MIM) device configuration using Pt as the bottom and top electrodes. The film capacitance (C_p) and dissipation factor ($\tan\delta$) were measured with an HP 4194A impedance/gain analyzer. The insulating properties of the films were evaluated via I-V measurements using a HP 4140B semiconductor test system from 0.1-1.0 MV/cm.

Both the buffer layer (Ta_2O_5/Si) and coplanar heterostructure device configuration (BST/Ta_2O_5/Si) were characterized for integration integrity via structural, microstructural and interfacial characterization techniques. Specifically, film crystallinity was assessed via glancing angle x-ray diffraction (GAXRD) using a Rigaku diffractometer with CuKa radiation at 40 kV. Cross-sectional film microstructure was examined using a Hitachi S4500 field emission scanning electron microscope (FESEM). The surface morphology of the constituent films and the composite heterostructure was assessed by a Digital Instrument's Dimension 3000 atomic force microscope (AFM) using tapping mode. Film residual stress was characterized via laser reflection curvature technique using a Tencor system. Rutherford backscattering spectrometry (RBS) was employed to assess film elemental composition, areal thickness, and film-substrate interface quality. The RBS measurements were obtained using 2 MeV He^+ ion beams from an

NEC 5SDH-2 tandem positive ion accelerator. All spectra were fit and interpreted using the program RUMP. [4]

RESULTS AND DISCUSSION

The Ta_2O_5 films were fabricated by the MOSD technique using carboxylate-alkoxide precursors. The as-pyrolyzed films (at 350 °C) were amorphous in nature and post-deposition annealing was required to obtain fully developed crystalline films, increase the overall grain size of the film, and to remove film strain by filling oxygen vacancies. GAXRD analysis of the films annealed at temperatures ranging from 600 to 750 °C, in 50° increments, for 1 hour determined the optimum annealing temperature to be 750 °C. At this temperature the films were well-crystallized, single phase, orthorhombic in structure with a preferential (200) orientation. The anneal-optimized Ta_2O_5 thin films were characterized for dielectric and insulating properties in the metal-insulator-metal MIM configuration. Figure 2 shows the dielectric permittivity and dissipation factor of the 100 nm thick Ta_2O_5 thin film as a function of frequency. The small signal dielectric constant and dissipation factor at a frequency of 100 kHz were 51.8 and 0.008, respectively. The permittivity and loss factor showed no appreciable dispersion with frequency up to about 1 MHz indicating that the values were not masked by any surface layer effects or electrode barrier effects in this frequency range.

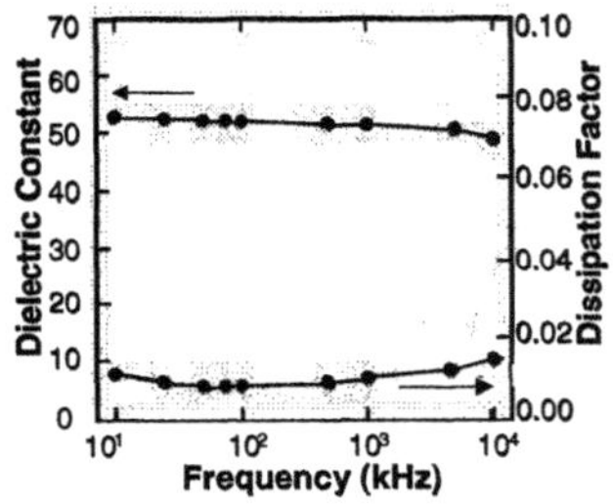

Figure 2. Dielectric constant and dissipation factor as a function of frequency for the 750 °C annealed Ta_2O_5 thin film.

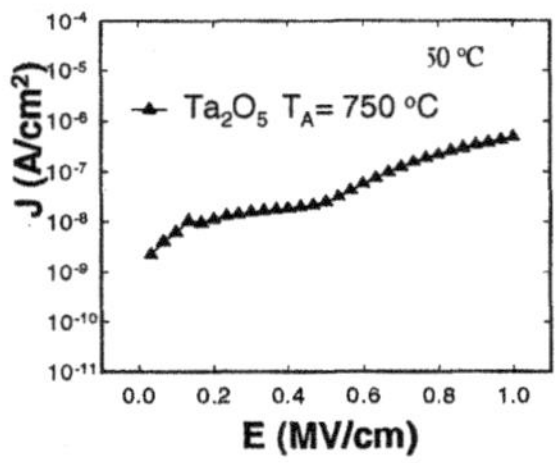

Figure 3. Leakage current density as a function of applied electric field for the 750 °C annealed Ta_2O_5 thin films.

Figure 3 displays the leakage current density for the anneal-optimized Ta_2O_5 thin film as a function of applied electric field. Leakage current is one of the limiting characteristics of a dielectric material for tunable device applications. The leakage current density of the thin film as a function of applied electric field is an electrical measure of the quality and reliability of a dielectric film.[1,5] Elevated leakage current/low resistivity values suggest poor device reliability and ultimate device failure/breakdown. The low leakage current or enhanced film resistivity value (2.0×10^{12} Ω-cm at 1MV/cm) observed in the present films demonstrates the completeness of phase formation and oxidation of the Ta_2O_5 MOSD fabricated thin films. The leakage current was observed to increase with applied electric field (Figure 3). It has been suggested that at high fields, grain boundary conduction may be larger than the grain conductivity setting large tunneling currents through these grain boundary layers. However even at these large field strengths the MOSD fabricated Ta_2O_5 films possessed excellent leakage characteristics.

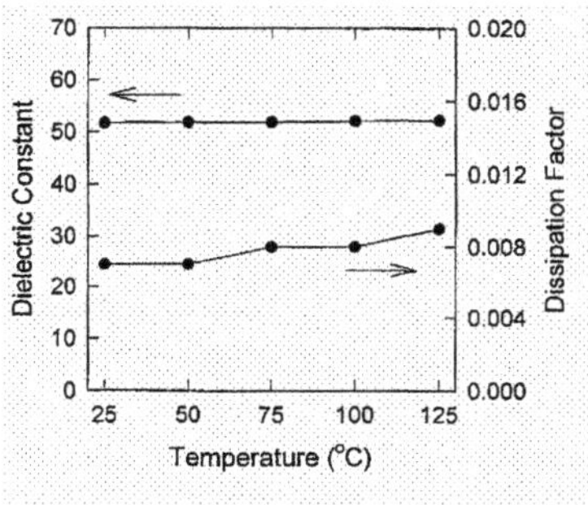

Figure 4. Permittivity and dissipation factor of the Ta_2O_5 thin film buffer layer as a function of measurement temperature from 25 to 125 °C.

In view of the fact that the Ta_2O_5 thin film will serve as an integration layer within the phase shifting element of the antenna its performance must not to perturb the phase shifter's in-service device operation. Thus, for Ta_2O_5 to be of practical use as an integration buffer layer in phase shifting device elements it must possess minimal dielectric dispersion with both temperature and applied bias. For example, an antenna/communication system must perform equally as well in the frigid temperatures of Russia as it does in the deserts of Kuwait, i.e. the antenna performance must not be modified/degraded by environmental ambient conditions such as temperature. It is well established that temperature sensitivity of the phase shifter's tunable (BST film) and/or passive (Ta_2O_5) constituents results in severe performance shortfalls.[6,7] Such performance shortfalls, namely beam pointing errors and device-to-device phase shift and/or insertion loss variations, are resultant of the highly non-linear temperature response of the constituent dielectric film(s), which make up the phase shifter. Figure 4 displays the dielectric dispersion of the anneal-optimized Ta_2O_5 thin films as a function of temperature from 25-125 °C. Both the permittivity and dissipation factor were relatively unchanged with measurement temperature in the range of 25 to 125 °C indicating excellent temperature stability of the buffer layer films. The temperature coefficient of capacitance (TCK) was calculated using the equation $\Delta C/(C_0 \Delta T)$, where ΔC is the change in capacitance relative to capacitance C_0 at 25 °C and ΔT is the change in temperature relative to 25 °C. The 750 °C anneal optimized films exhibited a low TCK of +114 ppm/°C confirming good reliability and integration integrity of the MOSD fabricated Ta_2O_5 films for buffer layers in phase shifting device elements.

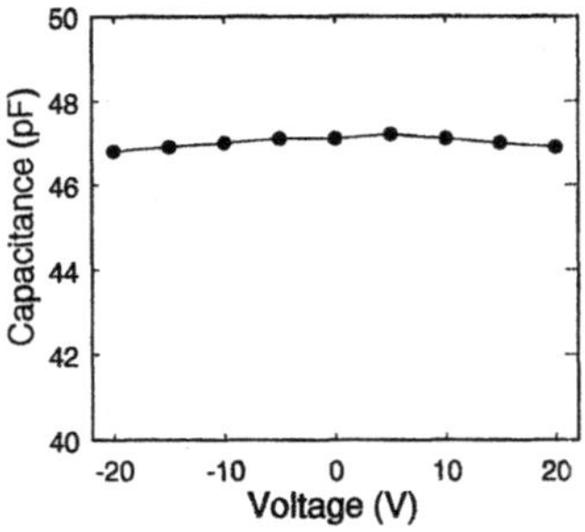

Figure 5. C-V characteristics of 750 ^{o}C annealed Ta_2O_5 thin films.

In addition to temperature stability, the buffer layer must not be tunable, i.e. it should be passive and exhibit good bias stability characteristics. Poor bias stability of the buffer layer would interfere with the tunability of the active material in the device and in turn would result in extreme unpredictable and uncorrectable errors in antenna phase shift. Such unexpected phase shift modifications would disable the antenna's ability to receive and/or transmit the information and ultimately result in communication failure. Therefore, it is necessary for the passive buffer layer component (Ta_2O_5) of the phase shifting element to be bias insensitive. The bias stability of the buffer layer film was determined via C-V measurements (Figure 5). The C-V measurements were obtained by applying a small ac signal of 10 mV amplitude and 100 kHz V across a Pt/ Ta_2O_5/Pt capacitor device structure, while the direct current electric field was swept from negative bias to positive bias. The change in film capacitance was determined to be 1.41% for the anneal optimized Ta_2O_5 films at an applied electric field of 1 MV/cm. The loss factor (not shown in Figure 5) also showed excellent bias stability, and was determined to be less than 1% up to 1 MV/cm.

In conjunction with the dielectric properties discussed above there are other materials properties, which must be satisfied for Ta_2O_5 to be a successful candidate as a buffer layer in phase shifting devices. For a buffer layer film to be monolithically integrated between the Si substrate and the overlying active thin film to create a bilayer heterostructure, the underling film (Ta_2O_5) must possess a pristine, surface morphology. In other words, a smooth, continuous, uniform, defect free surface morphology of the Ta_2O_5 film is critical if it is to serve as the "pseudo-substrate" for the growth of the overlying active layer, BST based thin film, in a heterostructure coplanar device configuration. If the Ta_2O_5 film surface is not smooth a rough BST film-Ta_2O_5 interface will result, and this interfacial roughness will in turn promote a rough BST film surface. In a coplanar microwave device design the top surface of the BST film must be metallized in order to fabricate the electrodes, i.e., the ground and center conductors of the device (Figure 1). There are losses associated with the electrode metallization and the quality of the film-metal electrode interface, which augment to the overall loss of the device. Minimization of film-metal interfacial roughness serves to reduce the conductor loss and improves the overall device loss for microwave frequency operation. Thus, the quality of the conductor-BST film interface is critical, and is strongly dependent on the nature of the buffer layer's surface morphology. Plan-view FESEM and AFM analyses were utilized to evaluate the 750 °C annealed Ta_2O_5 film surface for feasibility of promoting the growth of a continuous BST film with a uniform/smooth defect free surface morphology. The plan-view FESEM and AFM images of the 750 °C annealed Ta_2O_5 films on n^+-Si substrates are displayed in Figure 6. The micrographs show that the Ta_2O_5 films exhibited a smooth uniform microstructure with no cracks or defects observed. Furthermore, the microscopy results demonstrate that the Ta_2O_5 films possessed a dense microstructure composed of 200 nm domains, which encase a fine grain nano-crystallite microstructure or subgrains, with a grain size on the order of 25 nm. This domain-type microstructure is more pronounced/visible in the plan-view FESEM micrographs of the film surfaces (Figure 6a). The average surface roughness of the 750 °C post-deposition annealed films, as quantified by AFM, was found to be <1 nm. Thus, the Ta_2O_5 film surface properties, i.e., extreme uniformity, smoothness, and defect free nature, appear to be well suited for the BST overgrowth.

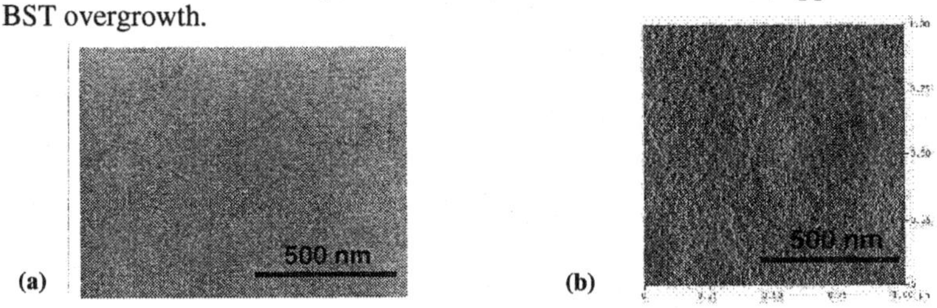

Figure 6. Surface morphology of the 750 °C annealed Ta_2O_5 thin films; (a) plan-view FESEM and (b) plan view AFM images.

Subsequent to the optimization of the Ta_2O_5 buffer layer film the challenge of integrating the buffer layer with the active BST thin film was addressed. To accomplish the integration task a dual anneal processing protocol was performed and optimized. In the dual anneal process protocol the two material depositions were crystallized by employing separate annealing crystallization steps for each film. Figure 7 displays the x-ray diffraction pattern of the dual anneal process protocol for the BST/Ta_2O_5 heterostructure on n^+-Si substrate. The x-ray diffraction results demonstrated that the films were well crystallized and that the appropriate crystalline phases of each material were present within the heterostructure. As such, the diffraction peaks form each layer of the heterostructure were easily identified and attributed to each of the constituent layers of the heterostructure with no evidence of secondary phases.

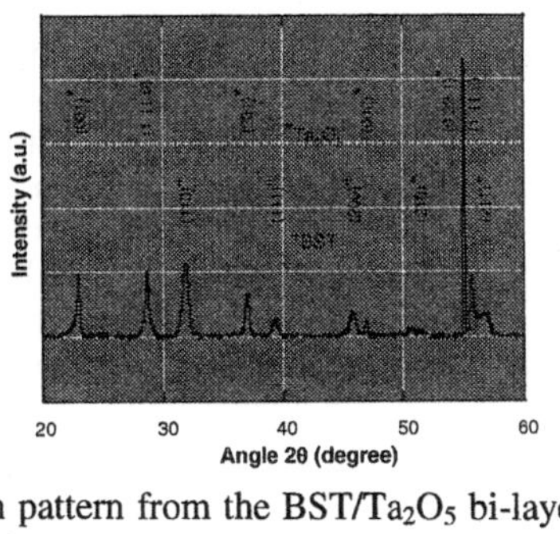

Figure 7. X-ray diffraction pattern from the BST/Ta_2O_5 bi-layer hetrostructure on Si fabricated

via the dual anneal process protocol. The Ta_2O_5 x-ray peaks are denoted with stars (*) and the x-ray peaks from the BST are denoted by crosses (+).

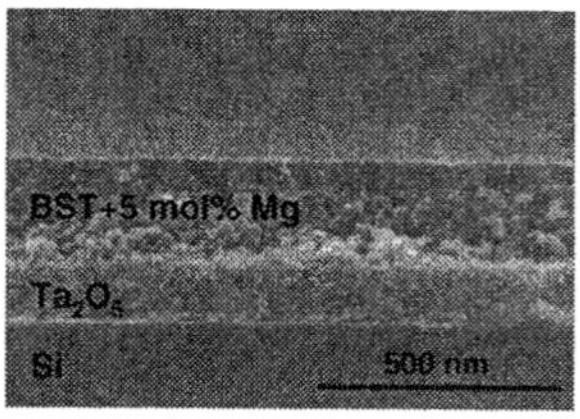

Figure 8. FESEM micrograph showing the cross-sectional microstructure and interface of the BST/Ta_2O_5 bi-layer film heterostructure fabricated via the dual anneal process protocol.

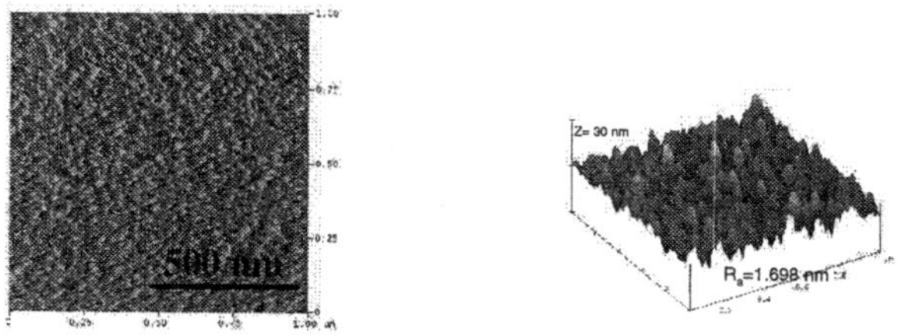

(a) (b)

Figure 9. AFM data for the BST/Ta_2O_5 bi-layer film heterostructure fabricated via the dual anneal process protocol: (a) plan view and (b) three dimensional AFM images.

The cross-sectional FESEM micrograph of the dual anneal heterostructure, displayed in Figure 8, demonstrated that the constituent films possessed a dense well-crystallized microstructure with uniform thickness. The FESEM micrograph shows a distinct structural delineation between the bilayer films and between the buffer layer film and the Si substrate. No amorphous layer or voiding/defects were observed at the interfaces. In addition, the surface morphology (Figure 9) of this heterostructure exhibited a dense microstructure with no cracks or defects observed. The surface roughness (Figure 9b) as quantified by AFM, was ~ 1.7 nm. The excellent crystallinity, defect free structurally abrupt interface in concert with an extremely smooth, uniform and defect free surface morphology bodes well for the integration suitability of this heterostructure fabricated via the dual anneal process protocol.

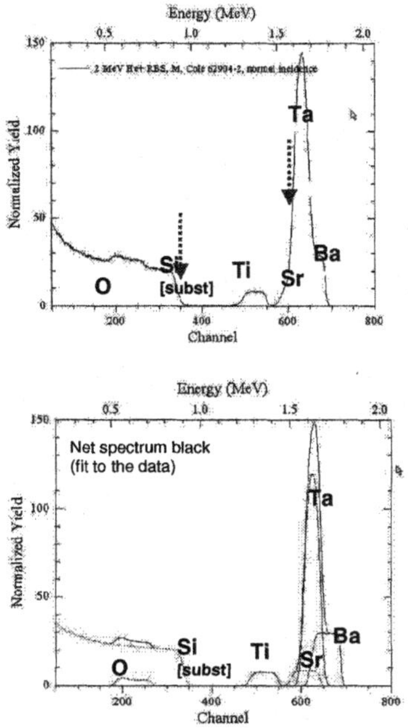

Figure 10. The RBS spectra of the dual anneal process protocol heterostructure: (a) the BST/Ta_2O_5/Si heterostructure data and (b) the BST/Ta_2O_5/Si simulation via RUMP.

For integrated thin-film device applications ideal performance of a coplanar heterosturctural design is best achieved by minimizing chemical imperfections such as variations in stoichiometry of the material. Such stoichiometric variations arise from elemental interdiffusion between the bilayer films and/or between the films and Si substrate. Rutherford backscatter spectroscopy (RBS) analysis was employed to access the prospect of elemental diffusion, and the interfacial quality of the dual anneal process protocol fabricated heterostructure. Specifically, composition, thickness, and interface quality of the constituent bilayer films were determined by comparing the experimental data to a simulation generated by the RUMP RBS analysis package.[4] The simulated spectrum was derived for an idealized BST/Ta_2O_5/Si heterostructure with no occurrence of elemental interdiffusion within the device structure. Figure 10 displays the RBS spectra for both the experimental data and RUMP simulation. A quantitative comparison of the experimental data with the simulated RBS results revealed no measurable deviation in peak shape and position between the two spectra. From the experimental RBS data (Figure 10a) it was observed that the low energy (back edges) edges of the Ba, Sr, Ti, and Ta peaks (from the bilayer film), and the high energy (front edge) edge of the Si signal from the substrate were sharp/abrupt. The fact that the experimental data is virtually identical to that of the simulation, combined with the peak edge sharpness, indicates that there was negligible interdiffusion within the heterostrucrure fabricated via the dual anneal process protocol. The RBS analyses confirmed the BST active thin film to be stoichiometric, i.e. the Ba:Sr ratio was confirmed to be 61.5:38.5. However, the stoichiometry of the Ta_2O_5 buffer layer was found to be $Ta_2O_{5.5}$, that is, the film contained about 10% excess oxygen. This excess oxygen may be due to absorbed water or contamination by low–Z oxides. The film thickness of the Ta_2O_5 and BST layers were determined to be ~100 and 200 nm, respectively, which is in agreement with the cross-sectional FESEM studies.

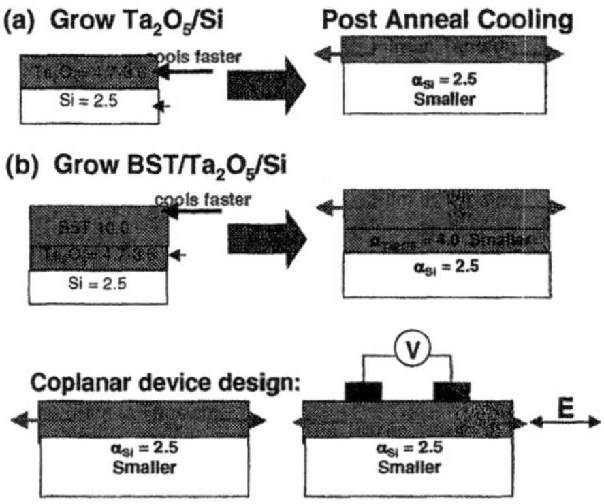

(c)

Figure 11. Schematic illustrating (a & b) the resultant residual stress due to the effect of CTE mismatch between integrated materials and post-anneal cooling, and (c) the films residual stress depicted in the coplanar device configuration under bias.

Our experimental results suggest that BST can be successfully integrated with affordable Si substrates utilizing a Ta_2O_5 buffer layer and a dual anneal process protocol. However, a major drawback of this device configuration, i.e. coplanar device design, is its lower tunability with respect to that of a parallel plate/vatactor device design.[6,8,9] We suggest that this device design performance short fall is mitigated by residual stress modification, which is inherent to our BST/Ta_2O_5/Si heterostructure design. Figure 11 (a & b) illustrates this residual stress modification which is resultant from the film CTE integration-mismatch and post-anneal cooling. Since the CTE of Si ($2.5\times10^{-6}\ ^{o}C^{-1}$) is much less than that of the Ta_2O_5 buffer layer ($4.7\ \times10^{-6}\ ^{o}C^{-1}$), subsequent to annealing the film will cool such that the Ta_2O_5 is in tension.[10] The residual stress of the 750 °C annealed Ta_2O_5 film on Si, measured via laser reflective curvature analysis, and was observed to be 21.9 MPa. The tensile nature of the Ta_2O_5 film is due to the CTE mismatch, post-anneal cooling, and the substrate effect. Specifically, as the Ta_2O_5/Si cools the film shrinks slower than the substrate thus an external force (CTE mismatch-substrate effect) causes the film to be in tension. Similarly, after the BST is deposited over the annealed Ta_2O_5 buffer layer and post-anneal cooled, it too will be in tension. In other words, the fact that the

CTE of Ta_2O_5 (4.7 $x10^{-6}$ $^oC^{-1}$) and the Si substrate ($2.5x10^{-6}$ $^oC^{-1}$) are significantly smaller than that of the BST film (10.5 $x10^{-6}$ $^oC^{-1}$), upon cooling the BST film will shrink slower than that of both the buffer layer and the substrate leaving the BST film in tension. The measured residual stress of the heterostructure bilayer (237.4 MPa) strongly supported this assertion. We suggest that this stress modification of the active material, via the integration with the Ta_2O_5 buffer layer and Si substrate, promotes enhanced tunability in the coplanar device design. In theory, in a coplanar device design, if the active thin film (BST) is under tension it is aligned parallel with the applied electric field and this configuration serves to promote the polarization of the electric dipoles (Figure 15c). In contract, an active film in compression constrains the polarization. Thus, this enhanced polarization results in a high capacitance density, which in turn promotes enhanced device tunability.

CONCLUSIONS

This investigation demonstrated the feasibility of utilizing MOSD fabricated Ta_2O_5 as a buffer layer film to promote the integration of BST with affordable large area silicon substrates for tunable device applications. The anneal-optimized Ta_2O_5 based thin film possessed excellent material properties, namely, an enhanced dielectric constant (ε_r =45.6), low dielectric loss (tan δ=0.006), high film resistivity ($\rho=10^{12}$ Ω-cm at E=1 MV/cm), excellent temperature stability (TCK of 52 ppm/^{o}C), and excellent bias stability of capacitance (~1.41% at 1 MV/cm). The integration of the BST/Ta_2O_5 bilayer heterostructure on Si substrate (in the coplanar device configuration), was demonstrated via the dual anneal process protocol. The defect free, structurally abrupt bilayer and bilayer film-substrate interface; combined with the smooth, uniform and defect free surface morphology of the dual anneal fabricated heterostructure suggest this process protocol to be an excellent method for realizing the successful monolithic integration of BST with affordable Si substrates in a coplanar device design. Our results suggest that it is the combination of material design and anneal process protocol, which promoted enhanced device tunability comparable to that of varactor devices.

REFERENCES

[1] M.W. Cole, P.C. Joshi, M.H. Ervin, M.C. Wood, and R.L. Pfeffer, "The Influence of Mg Doping on the Materials Properties of $Ba_{1-x}Sr_xTiO_3$ Thin Films For Tunable Device Applications", *Thin Solid Films*, **374**, 34-41 (2000).

[2] M.W. Cole, W.D. Nothwang, C. Hubbard, E. Ngo, and M.H. Ervin, "Low Dilectric Loss And Enhanced Tunability of $Ba_{1-x}Sr_xTio_3$ Thin Films Via Material Compositional Design and Optimized Film Processing Methods", *J. Appl. Phys.*, **93**, 9218-9225 (2003).

[3] S. Kim, B. Ahn, S. Cho, Ahn, J.Y. Lee, J.S. Chun, and W. Lee, Jpn. J., "Wafer Spin Etch Cleaning", *Appl. Phys., Part I*, **33**, 6691-6699 (1994).

[4] Doolittle, L.R., "Algorithms For Rapid Simulation of RBS Spectra:, *Nucl. Inst. Meth.*, **B15**, 334-351 (1985).

[5] S. Saha, and K.B. Krupanidhi, "Impact of Microstructure on the Electrical Stress Induced Effects of Pulsed Laser Ablated (Ba,Sr)TiO_3 Thin Films", *J. Appl. Phys.,* **87**, 3056- 3062 (2000).

[6] S.A. Wolf, and D. Treger, "Frequency Agile Materials For Electronics (FAME)-Progress In The DARPA Program", *Integr. Ferroelectrics*, **42**, 39-55 (2002).

[7] Coryell, L., "Private Communication", U.S. Army Communications and Electronics Research and Development Center, Ft. Monmouth, NJ (2003).

[8] R.G. Geyer, and M.W. Cole, "Microwave Properties of Acceptor-Doped Barium Strontium Titanate Thin Films For Tunable Electronic Devices", *Ceramic Transactions: Ceramic Materials and Multilayer Electronic Devices*, **150**, 229 – 244 (2004).

[9] J.S. Horwitz, W. Chang, A.C. Carter, J.M. Pond, S.W. Kirchoefer, D.B. Chrisey, J. Levy, C. Hubert, "Structure/Property Relationships in Ferroelectric Thin Films For Frequency Agile Microwave Electronics", *Integr. Ferroelectrics*, **22**, 279-289 (1998).

[10] C.-L. Tien, C.-C. Jaing, C. Lee, and K.-P. Chuang, "Simultaneous Determination of The Thermal Expansion Coefficient And The Elastic Modulus of Ta_2O_5 Thin Films Using Phase Shifting Interferometry", *Journal of Modern Optics*, **47**, 1681-1691(2002).

Author Index

Lightning Source UK Ltd.
Milton Keynes UK
UKOW06f2014131116
287480UK00001B/81/P

9 781574 982626